Aquatic Ecosystem FOUNDATION

October 2009

Dear Reader:

Thank you for your interest in aquatic plant management. The Aquatic Ecosystem Restoration Foundation (AERF) is pleased to bring you *Biology and Control of Aquatic Plants: A Best Management Practices Handbook*.

The mission of the AERF, a not for profit foundation, is to support research and development which provides strategies and techniques for the environmentally and scientifically sound management, conservation and restoration of aquatic ecosystems. One of the ways the Foundation accomplishes the mission is by providing information to the public on the benefits of conserving aquatic ecosystems. The handbook has been one of the most successful ways of distributing information to the public regarding aquatic plant management. The first edition of this handbook became one of the most widely read and used references in the aquatic plant management community. This second edition has been specifically designed with the water resource manager, water management association, homeowners and customers and operators of aquatic plant management companies and districts in mind. It is not intended to provide the answers to every question, but it should provide basic scientifically sound information to assist decision-makers.

The authors, editors and contributors reflect the best the aquatic plant management industry has to offer. They gave generously of their time and talent in the production of this document and they deserve all the praise and thanks that can be garnered. Not only have they prepared the chapters and appendices, they are available to all interested parties to provide clarification and additional information as warranted. These scientists, professors, aquatic plant managers and government officials have created a document that surely will be the most widely read and circulated handbook produced to date. Thank you all.

The production of this document has been made possible through the generosity of members of the Foundation. My thanks and appreciation to these faithful supporters who continue to underwrite what has been an effort to provide the very best handbook possible.

I hope you find this handbook to be helpful and informative. A downloadable version is on the AERF website at www.Aquatics.org along with other useful information and links. Consider becoming a member of the Foundation and supporting educational projects and other ecosystem restoration efforts across the country.

Carlton R. Layne
Executive Director

Contributors

Marc Bellaud (mbellaud@aquaticcontroltech.com) - Chapters 6 and 13.10; co-editor
Aquatic Control Technology Inc.
11 John Road
Sutton MA 01590-2509

Douglas Colle (dcolle@ufl.edu) - Chapter 10
University of Florida Fisheries and Aquatic Sciences Program
School of Forest Resources and Conservation
7922 NW 71st Street
Gainesville FL 32653

James Cuda (jcuda@ufl.edu) - Chapters 5, 8 and 9
University of Florida Department of Entomology and Nematology
Box 110620
Gainesville FL 32611-0620

Eric Dibble (edibble@cfr.msstate.edu) - Chapter 2
Mississippi State University Department of Wildlife and Fisheries
Box 9690
Mississippi State MS 39762-9690

Kurt Getsinger (Kurt.D.Getsinger@usace.army.mil) - Chapter 3
US Army Engineer Research and Development Center
3909 Halls Ferry Road
Vicksburg MS 39180-6199

Lyn Gettys (lgettys@ufl.edu) - Chapter 13.5; co-editor
University of Florida Center for Aquatic and Invasive Plants
7922 NW 71st Street
Gainesville FL 32653

William Haller (whaller@ufl.edu) - Chapters 7, 13 introduction, 13.1, Appendix B, F; co-editor
University of Florida Center for Aquatic and Invasive Plants
7922 NW 71st Street
Gainesville FL 32653

Mark Hoyer (mvhoyer@ufl.edu) - Chapter 4
University of Florida Fisheries and Aquatic Sciences Program
School of Forest Resources and Conservation
7922 NW 71st Street
Gainesville FL 32653

Robert Johnson (rlj5@cornell.edu) - Chapter 13.6
Cornell University Research Ponds Facility
E140 Corson Hall
Ithaca NY 14853

Scott Kishbaugh (sakishba@gw.dec.state.ny.us) - Chapter 13.3
Bureau of Water Assessment and Management
New York State Department of Environmental Conservation, Division of Water
625 Broadway, 4th Floor
Albany NY 12233-3502

Carlton Layne (layn1111@bellsouth.net) - Appendix A
US Environmental Protection Agency (retired)
3272 Sherman Ridge Drive
Marietta GA 30064

Carole Lembi (lembi@purdue.edu) - Chapter 12
Purdue University Department of Botany and Plant Pathology
915 W. State Street
West Lafayette IN 47907-2054

John Madsen (jmadsen@gri.msstate.edu) - Chapters 1, 13.2, Appendix D
Mississippi State University Geosystems Research Institute
Box 9652
Mississippi State MS 39762-9652

Bernalyn McGaughey (bmcgaughey@complianceservices.com) - Appendix C
Compliance Services International
7501 Bridgeport Way West
Lakewood WA 98499

Linda Nelson (Linda.S.Nelson@usace.army.mil) - Chapter 13.4
US Army Engineer Research and Development Center
3909 Halls Ferry Road
Vicksburg MS 39180-6199

Michael Netherland (mdnether@ufl.edu) - Chapter 11, Appendix E
US Army Engineer Research and Development Center
7922 NW 71st Street
Gainesville FL 32653

Toni Pennington (toni.pennington@tetratech.com) - Chapter 13.8
Tetra Tech, Inc.
1020 SW Taylor Street, Suite 530
Portland OR 97205

David Petty (dpetty@ndrsite.com) - Appendix F
NDR Research
710 Hanna Street
Plainfield IN 46168

Jeffrey Schardt (jeff.schardt@myfwc.com) - Appendix E
Florida Fish and Wildlife Conservation Commission
2600 Blair Stone Road M.S. 3500
Tallahassee FL 32399

Donald Stubbs (donald271@verizon.net) - Appendix A
US Environmental Protection Agency (retired)
2301 Home Farm Court
Gambrils MD 21054

Ryan Wersal (rwersal@gri.msstate.edu) - Chapter 3
Mississippi State University Geosystems Research Institute
Box 9652
Mississippi State MS 39762-9652

Jack Whetstone (jwhtstn@clemson.edu) - Chapter 13.9
Clemson University Baruch Institute of Coastal Ecology and Forest Science
PO Box 596
Georgetown SC 29442

Thomas Woolf (twoolf@agri.idaho.gov) - Chapter 13.7
Idaho State Department of Agriculture
2270 Old Penitentiary Road
Boise ID 83701

Please cite this document using the following format:
Gettys LA, WT Haller and M Bellaud, eds. 2009. Biology and control of aquatic plants: a best management practices handbook. Aquatic Ecosystem Restoration Foundation, Marietta GA. 210 pages.

Please cite chapters in this document using the following format:
Haller WT. 2009. Chapter 7: mechanical control of aquatic weeds, pp. 41-46. In: Biology and control of aquatic plants: a best management practices handbook (Gettys LA, WT Haller and M Bellaud, eds.). Aquatic Ecosystem Restoration Foundation, Marietta GA. 210 pages.

Aquatic Ecosystem Restoration FOUNDATION

Table of Contents

Chapter 1: Impact of Invasive Aquatic Plants on Aquatic Biology 1
John Madsen, Mississippi State University

Chapter 2: Impact of Invasive Aquatic Plants on Fish 9
Eric Dibble, Mississippi State University

Chapter 3: Impact of Invasive Aquatic Plants on Waterfowl 19
Ryan Wersal, Mississippi State University
Kurt Getsinger, US Army Engineer Research and Development Center

Chapter 4: Impact of Invasive Aquatic Plants on Aquatic Birds 25
Mark Hoyer, University of Florida

Chapter 5: Aquatic Plants, Mosquitoes and Public Health 31
James Cuda, University of Florida

Chapter 6: Cultural and Physical Control of Aquatic Weeds 35
Marc Bellaud, Aquatic Control Technology, Inc.

Chapter 7: Mechanical Control of Aquatic Weeds 41
William Haller, University of Florida

Chapter 8: Introduction to Biological Control of Aquatic Weeds 47
James Cuda, University of Florida

Chapter 9: Insects for Biocontrol of Aquatic Weeds 55
James Cuda, University of Florida

Chapter 10: Grass Carp for Biocontrol of Aquatic Weeds 61
Douglas Colle, University of Florida

Chapter 11: Chemical Control of Aquatic Weeds 65
Michael Netherland, US Army Engineer Research and Development Center

Chapter 12: The Biology and Management of Algae 79
Carole Lembi, Purdue University

Chapter 13: Introduction to the Plant Monographs 87
William Haller, University of Florida

    Chapter 13.1: Hydrilla 89
    William Haller, University of Florida

    Chapter 13.2: Eurasian Watermilfoil 95
    John Madsen, Mississippi State University

Chapter 13.3: Waterchestnut 99
Scott Kishbaugh, New York State Department of Environmental Conservation

Chapter 13.4: Giant and Common Salvinia 105
Linda Nelson, US Army Engineer Research and Development Center

Chapter 13.5: Waterhyacinth 113
Lyn Gettys, University of Florida

Chapter 13.6: Purple Loosestrife 119
Robert Johnson, Cornell University

Chapter 13.7: Curlyleaf Pondweed 125
Thomas Woolf, Idaho State Department of Agriculture

Chapter 13.8: Egeria 129
Toni Pennington, Tetra Tech, Inc.

Chapter 13.9: Phragmites—Common Reed 135
Jack Whetstone, Clemson University

Chapter 13.10: Flowering Rush 141
Marc Bellaud, Aquatic Control Technology, Inc.

Appendix A: Requirements for Registration of Aquatic Herbicides 145
Carlton Layne, US EPA (retired)
Donald Stubbs, US EPA (retired)

Appendix B: Aquatic Herbicide Application Methods 151
William Haller, University of Florida

Appendix C: A Discussion to Address Your Concerns: Will Herbicides Hurt Me or My Lake? 157
Bernalyn McGaughey, Compliance Services International

Appendix D: Developing a Lake Management Plan 167
John Madsen, Mississippi State University

Appendix E: A Manager's Definition of Aquatic Plant Control 173
Michael Netherland, US Army Engineer Research and Development Center
Jeffrey Schardt, Florida Fish and Wildlife Conservation Commission

Appendix F: Miscellaneous Information 181
William Haller, University of Florida
David Petty, NDR Research

Glossary 189

John D. Madsen
Mississippi State University, Mississippi State MS
jmadsen@gri.msstate.edu

Chapter 1

Impact of Invasive Aquatic Plants on Aquatic Biology

Introduction

Aquatic plants play an important role in aquatic systems worldwide because they provide food and habitat to fish, wildlife and aquatic organisms. Plants stabilize sediments, improve water clarity and add diversity to the shallow areas of lakes. Unfortunately, nonnative plants that are introduced to new habitats often become a nuisance by hindering human uses of water and threaten the structure and function of diverse native aquatic ecosystems. Significant resources are often expended to manage infestations of aquatic weeds because unchecked growth of these invasive species often interferes with use of water, increases the risk of flooding and results in conditions that threaten public health.

Types of aquatic plants

Aquatic plants grow partially or completely in water. Macrophytic plants are large enough to be seen with the naked eye (as compared to phytoplankton, which are tiny and can only be identified with a microscope) and are found in the shallow zones of lakes or rivers. This shallow zone is called the littoral zone and is the area where sufficient light penetrates to the bottom to support the growth of plants. Plants that grow in littoral zones are divided into three groups. Emergent plants inhabit the shallowest water and are rooted in the sediment with their leaves extending above the water's surface. Representative species of emergent plants include bulrush, cattail and arrowhead. Floating-leaved plants grow at intermediate depths. Some floating-leaved species are rooted in the sediment, but others are free-floating with roots that hang unanchored in the water column. The leaves of floating-leaved plants float more or less flat on the surface of the water. Waterlily and spatterdock are floating-leaved species, whereas waterhyacinth (Chapter 13.5) and waterlettuce are free-floating plants. Submersed plants are rooted in the sediment and inhabit the deepest fringe of the littoral zone where light penetration is sufficient to support growth of the plant. Submersed plants grow up through the water column and the growth of most submersed species occurs entirely within the water column, with no plant parts emerging from the water. Submersed species include hydrilla (Chapter 13.1), curlyleaf pondweed (Chapter 13.7), egeria (Chapter 13.8) and vallisneria.

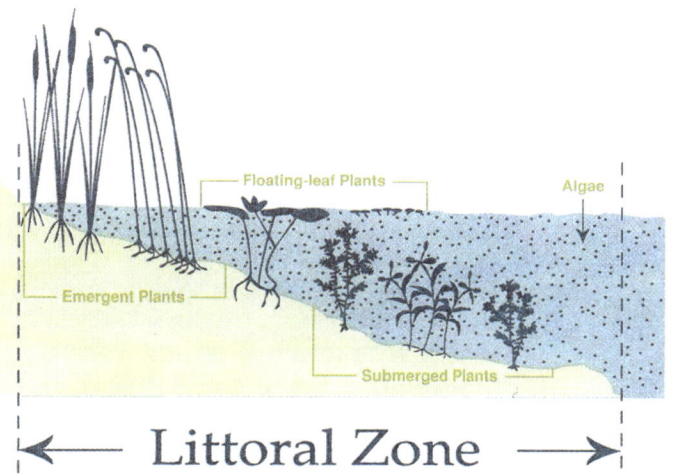

Algae also grow in lakes and provide the basis of the food chain. The smallest algae are called phytoplankton and are microscopic cells that grow suspended in the water column throughout the lake. Dense growth of phytoplankton may make water appear green, but even the "cleanest" lake with no green coloration has phytoplankton suspended in the water. Filamentous algae grow as chains of cells and may form large strings or mats. Some filamentous algae are free-floating and grow suspended in the water column, but other species grow attached to plants or the bottom of the lake. Macroscopic or macrophytic algae are large green organisms that look like submersed plants, but are actually algae (Chapter 12).

What aquatic plants need

Plants have simple needs in order to grow and thrive – they require carbon dioxide, oxygen, nutrients, water and light. Plants use light energy, water and carbon dioxide to synthesize carbohydrates and release oxygen into the environment during photosynthesis. Animals use both the carbohydrates and oxygen produced by plants during photosynthesis to survive, so without plants there would be no animal life. The nutrients required in the greatest quantity by plants are nitrogen and phosphorus, but a dozen or more other minerals are also needed to support plant growth. Plant cells use oxygen in the process of respiration just like animal cells, but this is often forgotten since plants produce more oxygen than they need for their own use.

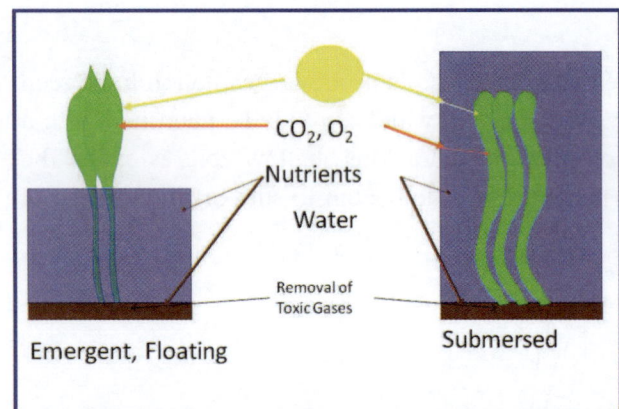

Aquatic plants inhabit an environment very favorable in one respect – most terrestrial plants must find sufficient water to survive. Aquatic plants are literally bathed in water, one of the primary requirements for plant growth. Since aquatic sediments are typically high in nitrogen and phosphorus, life might appear idyllic for aquatic plants. Once the leaves of emergent and floating–leaved plants rise above the water surface, they have a ready supply of carbon dioxide, oxygen and light. In addition, the leaves may act as a conduit for the ready disposal of toxic gases like methane and sulfur dioxide produced in the sediments surrounding plant roots. Given these factors, it is no surprise that emergent plants in fertile marshes are among the most productive ecosystems in the world.

Alas, life is not as easy for submersed plants. While submersed plants have easy access to the same pool of nutrients from the water and the sediment, the availability of light and carbon dioxide is significantly reduced since most submersed plants live completely under the water. Light must penetrate through the water column to reach submersed plants; therefore, much less light energy is available to them. Also, carbon dioxide must be extracted from the water, an environment in which carbon dioxide is present in much lower concentrations and diffuses much more slowly than in the air. As a result, submersed plants are much less productive than emergent and floating plants and the primary factors limiting their growth are the availability of light and carbon dioxide. Some highly productive plants have developed means to increase their access to light and carbon dioxide. For example, species such as hydrilla form dense canopies on the surface of the water, which allows them to capture light energy that is less available near the bottom of the water column. These productive (and often invasive) aquatic plants form dense colonies that interfere with human uses of the littoral areas, increase flooding risk and shade out plants – including most native species – that do not form canopies.

Lake ecology
Trophic state

Trophic state describes the overall productivity (amount of plants or algae) of a lake, which has implications for the biological, chemical and physical conditions of the lake. For example, aquatic animals use plants as a food source, so unproductive lakes do not support large populations of zooplankton, invertebrates, fish, birds, snakes and other animals. The trophic state of a lake is directly tied to the overall algal productivity of the lake and ranges from very unproductive to highly productive. Because phytoplankton typically control lake productivity, factors that increase algal productivity also increase the trophic state of the lake. Algal biomass in a lake is estimated by measuring the concentration of chlorophyll in the water; hence, lake chlorophyll concentration is a direct measure of lake trophic state.

Chlorophyll is directly related to phosphorus concentration in the lake, so phosphorus is also considered a direct measure of lake trophic state. Lake transparency is the most widely measured characteristic to determine trophic state because growth of algae increases water turbidity – high algal growth reduces water clarity, which suggests high productivity. Trophic state can be measured with a Secchi disk because most turbidity in lakes is caused by suspended algae. Since increased algal growth makes the water less transparent, Secchi disk depth is a measure of lake trophic state. Chlorophyll, phosphorus and Secchi disk depth are measured in different units. The Trophic State Index (TSI) employs equations that allow users to develop a single uniform number for trophic state based on any one of the three factors alone or on the average of all three factors (chlorophyll, total phosphorus or Secchi disk depth). This tool is useful to compare trophic state data collected by differing methods and has empowered hundreds of lay monitors to collect trophic state data using only a Secchi disk to estimate water clarity.

Four terms are commonly used to describe lake trophic state. <u>Oligotrophic</u> lakes are unproductive with low nutrients (phosphorus < 15 µg/L) and low algal productivity (chlorophyll < 3 µg/L). Transparency, as measured by the Secchi disk method, is greater than 13 feet. Oligotrophic lakes are typically well-oxygenated and often support cold-water fisheries in the northern US. <u>Mesotrophic</u> lakes are moderately productive, with intermediate levels of chlorophyll, nutrients, and water clarity. Mesotrophic lakes may support abundant populations of rooted aquatic plants and often have cool-water fisheries. <u>Eutrophic</u> lakes are highly productive, with high levels of phosphorus and chlorophyll. Water clarity is low and generally ranges from 3 to 8 feet as measured by the Secchi disk method. Eutrophic lakes may support bass fisheries but rarely have productive open-water fisheries. <u>Hypereutrophic</u> lakes have very high phosphorus and chlorophyll levels and water clarity is usually less than 3 feet. In most cases, hypereutrophic lakes are the result of nutrient loading from human activity in the watershed. Algal growth dominates in the lake and few or no rooted plants are present.

Trophic state	Chlorophyll concentration (μg/L)	Total phosphorus concentration (μg/L)	Water clarity (by Secchi Disk, in feet)	Trophic State Index	Description
Oligotrophic	< 3	<15	>13	<30	Very low productivity Clear water Well oxygenated Few plants and animals
Mesotrophic	3-7	15-25	8-13	40-50	Low to medium productivity Moderately clear water Abundant plant growth
Eutrophic	7-40	25-100	3-8	50-60	Medium to high productivity Fair water clarity Dense plant growth
Hypereutrophic	>40	>100	<3	>70	Very high productivity Poor water clarity Limited submersed plant growth, algae dominate

Studies of sediment cores from lakes across the US have verified that many lakes were naturally mesotrophic or eutrophic before Europeans settled in the US, which conflicts with the assumption that all "pristine" lakes are oligotrophic. The nutrient status or trophic state of lakes that are unaffected by human activity is a function of the watershed and its geology. That being said, human activity that causes nutrient runoff into lakes can shift a lake to a higher trophic state, which alters many biological and chemical attributes of the lake. There are many examples of pollution-degraded lakes, but the water quality of many lakes has improved since the passage of the first Clean Water Act and these lakes are returning to their historic water quality levels due to efforts to restore our waterways.

Productivity in lakes

As mentioned above, algae and macrophytic plants are the basis for lake productivity. Plants take up nutrients, water and carbon dioxide from the environment and use light energy to produce carbohydrates and sugars, with oxygen as a byproduct. Herbivores such as crustaceans and insects consume aquatic plants and use energy from the plants to grow. Forage fish such as minnows and bluegill consume these herbivores and use energy from the herbivores to grow. Fish-eating fish such as trout, bass, pike and walleye eat these forage fish and use energy from the forage fish to grow. Because each level of this feeding system is based on

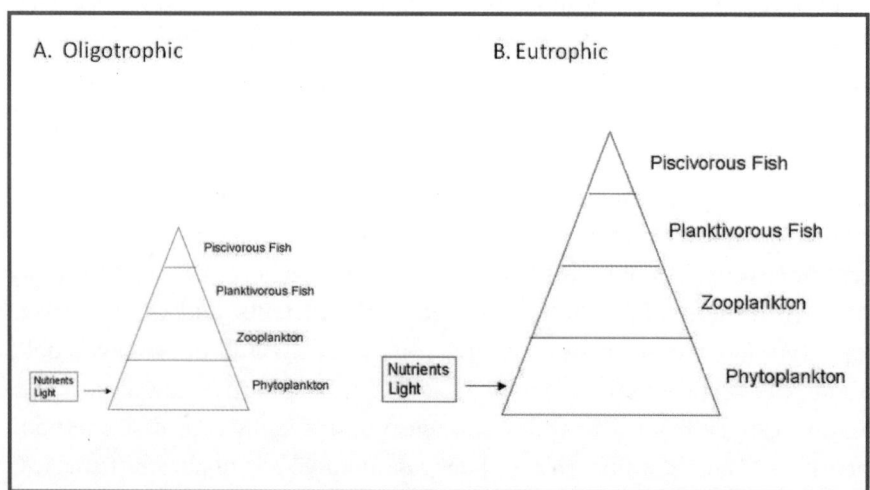

the energy of the level below it, this system is often described as a food pyramid. Oligotrophic lakes with few nutrients and little plant production have small pyramids, whereas eutrophic lakes with much higher nutrient concentrations, more total plant growth (algae and rooted plants) and more fish have larger pyramids. This relationship has been recognized by the aquaculture industry and fertilizer is frequently added to production ponds to increase fisheries productivity. However, changes in water quality can increase populations of undesirable fish as well as populations of more desirable species in reservoirs and in natural systems.

Food chains in lakes
A food chain is a depiction of what various organisms in an ecosystem consume. Food chains begin with algae and plants, which are followed by herbivores, small forage fish and finally by the top-level predator. There may be a hundred species in a lake, so the food chain is often simplified to include only the dominant species. Phytoplankton form the base of the food chain in a typical pelagic (open-water) zone. Phytoplankton are consumed by zooplankton (small crustaceans) that are suspended in the water. Zooplankton are in turn eaten by smaller fish such as yellow perch. Yellow perch are then consumed by the top predator such as walleye.

The food chain in the littoral zone is different. Some algae are present – both as phytoplankton and as algae growing on plant surfaces – but much of the food is derived from macrophytic plants. Most macrophytes are consumed only after they have died and partially decomposed into detritus. Detritus is eaten primarily by aquatic insects, invertebrates and larger crustaceans. These detritivores, which live on or near the lake bottom, are in turn consumed by the dominant littoral forage fish such as bluegill sunfish. Lastly, forage fish are consumed by the top predator such as largemouth bass.

Littoral and cold-water pelagic zone food chains are often isolated from each other and almost function as two separate ecosystems within the same lake. The substantial changes caused by shifts between these food chains are exemplified by the history of Lake St. Clair in Michigan. Lake St. Clair only looks small compared to the Great Lakes it lays between – Lakes Huron and Erie. In fact, it is a 430 square mile lake with a maximum depth of 30 feet, although over 90% of the lake is 12 feet deep or less. This shallow lake was very turbid before 1970, with a Secchi disk transparency of only 4 feet. Rooted plants grew in about 20% of the lake and Lake St. Clair was home to a world-class commercial and recreational open-water walleye and yellow perch fishery. Lake St. Clair was invaded in the 1980s by the zebra mussel, an invasive bivalve (clam) that filters water by consuming suspended phytoplankton and the nutrients associated with them. Zebra mussels filtered the water of Lake St. Clair so effectively that water transparency more than doubled a few years after their invasion. Rooted plants expanded to almost 80% of the lake due to increased light penetration and the fishery completely changed. Walleye and yellow perch can still be found, but the former open-water fishery is now used largely for recreational angling for largemouth bass, a typical littoral zone predator.

Aquatic plant communities
Native aquatic plant species tend to separate into depth zone bands (referred to as depth zonation), with a mix of species found in each depth zone. Submersed plants may be found in water as deep as 30 feet or more in oligotrophic lakes and distinct bands of vegetation are visible to the shoreline. Plants in oligotrophic lakes are adapted to low levels of nutrients and carbon dioxide. Light penetrates easily to 30 feet or more and light levels are not limiting, but plants are typically very short. Submersed aquatic mosses also grow at water depths of up to 200 feet in Crater Lake in Oregon. Plant diversity is often relatively low and native plants in oligotrophic lakes rarely form populations that are substantial enough to cause problems.

Depth zonation in mesotrophic lakes is likewise pronounced, with submersed plants growing in water as deep as 15 to 20 feet. Submersed plants may grow to reach the surface of the water, but this growth is typically localized and occurs in water that is less than 10 feet deep. Plant species diversity is usually at a maximum in mesotrophic lakes; numerous plant growth forms are present and result in a multilayered plant canopy. Light penetration may limit plant growth but plants grow at depths greater than in eutrophic lakes and the total amount of plant growth in mesotrophic lakes is often as high as in eutrophic systems. Nutrients rarely limit plant growth in mesotrophic systems and growth of aquatic species is almost completely dependent on light penetration. Residents living next to reservoirs and lakes often report changes in plant coverage from year to year; these changes are typical of dynamic mesotrophic systems and are usually the result of changes in light penetration.

Depth zonation in eutrophic lakes is much less pronounced, with plant growth typically occurring at maximum depths of only 12 to 15 feet. Plant abundance is high, but plant diversity is much lower than in mesotrophic lakes and erect and canopy-forming plants predominate because light is often limited due to growth of phytoplankton. Native plants often produce populations that are large enough to be nuisances, particularly in high-use areas such as boat ramps and swimming areas. Light strongly limits plant growth and canopy-forming plants have a distinct advantage over plants that do not form canopies.

Hypereutrophic lakes typically have poorly developed aquatic plant communities and plants rarely grow in water more than 6 feet deep. Some emergent and floating plants can be found, but submersed plant growth is greatly reduced and typically only canopy-forming species are able to establish. Plants that are able to colonize hypereutrophic lakes often grow to nuisance levels. High algal production results in dense blooms that intercept available light. As a result, plant diversity is low and the abundance of rooted plants is typically lower than in eutrophic lakes.

So what should a typical lake look like? Well, that depends. Without human-mediated nutrient loading from sewage treatment plants and runoff from fields and residential areas, hypereutrophic lakes would be rare occurrences. Therefore, the natural state of a typical lake would include a littoral zone dominated by aquatic plants. Even in eutrophic lakes, nuisance populations of native plants would likely be localized and would cause problems only when the plants interfere with recreational or other uses. However, the introduction of invasive exotic plants changes this dynamic, even in oligotrophic lakes.

Invasive plants

Invasive aquatic plants are generally defined as nonnative (from another geographic region, usually another continent) plant species that cause ecological and/or economic harm to a natural or managed ecosystem. Invasive aquatic plants often cause both economic and ecological harm.

As invasive plants expand in a new area, they suppress the growth of native plants and cause localized extinction of native species. For instance, when Eurasian watermilfoil (Chapter 13.2) invaded Lake George in New York, growth of this exotic species reduced the total number of species in a permanent research plot from 21 to 9 over a three-year period. Invasive plant species can invade a particular zone of the depth profile and suppress the native plant species that normally inhabit the area. Colonization by invasive species may be less damaging in oligotrophic lakes, because native plants can grow at much greater depths than invasive species. Native plants often persist in areas of mesotrophic lakes that are shallower and deeper than those colonized by invasive plants. Invasive plants dominate to the borders of eutrophic and hypereutrophic lakes, with native plants often confined to a shallow fringe around the lake.

Economic impacts	Ecological effects
Impair commercial navigation	Degrade water quality
Disrupt hydropower generation	Reduce species diversity
Increase flood frequency, duration and intensity	Suppress desirable native plants
Impair drinking water (taste and odor)	Increase extinction rate of rare, threatened and endangered species
Habitat for insect-borne disease vectors	Alter animal community interactions
Recreational navigation impairment	Increase detritus buildup
Interfere with safe swimming	Change sediment chemistry
Interfere with fishing	
Reduce property value	
Endanger human health, increase drowning risk	

Summary

Invasive plants reduce native plant growth and impede human uses of waters by forming dense surface canopies that shade out lower-growing native plants and interfere with water flow, boat traffic and fishing. Dense surface canopies also radically change the habitat quality for fish. Dense plant beds provide a place for small forage fish to hide and reduce the ability of predatory fish such as bass and northern pike to see their prey. This tends to lead to a large number of small, stunted forage fishes and poor production of game fishes (Chapter 2).

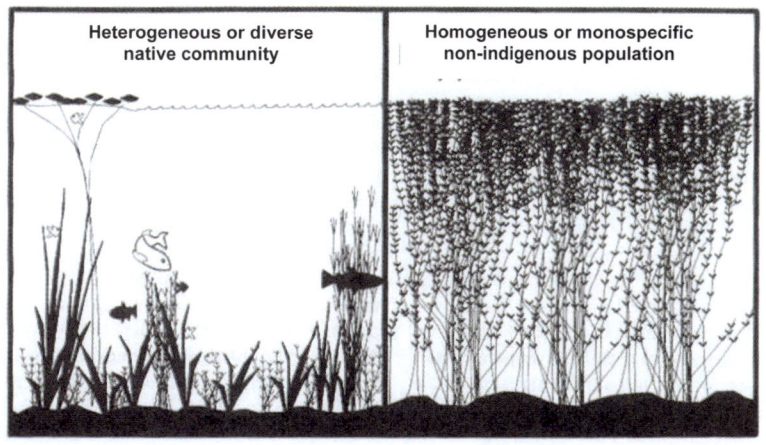

Invasive plants also reduce water quality. While the increased biomass and dense canopies formed by invasive species tend to increase water clarity, they also lead to increased organic sedimentation. The fate of all lakes over geological time is to progress from lakes to wetlands to marshes to upland areas as lakes fill with sediments due to erosion and accumulation of organic matter. Exotic plants are also significantly more productive than native species and increase the rate of nutrient loading in the system by utilizing nitrogen and phosphorus from the sediment. For example, curlyleaf pondweed has been implicated in increased internal nutrient loading in Midwestern lakes because the plants absorb nutrients from the sediments and grow throughout the spring and summer, then die and release the nutrients into the water. Water also becomes stagnant under dense plant canopies and suppresses or prevents oxygen recirculation. In addition, the amount of dissolved oxygen under dense plant canopies may be insufficient to support desirable fish species and may result in fish kills.

Many animal species are linked to specific native plant communities and the diversity of native communities provides a variety of habitats for aquatic insects and other fauna. Invasive plants reduce the diversity of native plant communities, which leads to a reduction in the diversity of both fish and aquatic insects. Therefore, invasive plants are harmful to the diversity and function of aquatic ecosystems and can have significant adverse impacts on water resources.

For more information:
- Plant growth form definition
http://www.dnr.state.mn.us/shorelandmgmt/apg/wheregrow.html
- Lake productivity
http://www.ecy.wa.gov/Programs/wq/plants/management/joysmanual/lakedata.html
- Lake food chains
http://www.wateronthweb.org/under/lakeecology/11_foodweb.html
- Lake trophic state
http://aquat1.ifas.ufl.edu/guide/trophstate.html
http://lakewatch.ifas.ufl.edu/circpdffolder/trophic2.pdf
- Secchi disk
http://dipin.kent.edu/secchi.htm
http://www.epa.gov/volunteer/lake/lakevolman.pdf
- Lake TSI (trophic state index)
http://dipin.kent.edu/tsi.htm
- Depth zonation of aquatic plants
http://www.niwa.co.nz/news-and-publications/publications/all/wa/10-1/submerged
- Invasive aquatic plants
http://aquat1.ifas.ufl.edu
http://www.dnr.state.mn.us/invasives/aquaticplants/index.html

Photo and illustration credits:
Page 1: Littoral zone; Minnesota Department of Natural Resources
Page 2: Aquatic plants illustration; John Madsen, Mississippi State University Geosystems Research Institute
Page 3: Secchi disk; Margaret Glenn, University of Florida Center for Aquatic and Invasive Plants
Page 4: Food pyramids; John Madsen, Mississippi State University Geosystems Research Institute
Page 7: Heterogeneous and homogeneous plant communities; Robert Doyle, Baylor University

Chapter 2

Eric Dibble
Mississippi State University, Mississippi State MS
edibble@cfr.msstate.edu

Impact of Invasive Aquatic Plants on Fish

Introduction

Many species of fish rely on aquatic plants at some point during their lives and often move to different habitats based on their growth stage. Young fish use the cover provided by aquatic vegetation to hide from predators and their diets may be dependent on algae and the microfauna (e.g., zooplankton, insects and larvae) that live on aquatic plants. Mature fish of some species move to more open waters to reduce foraging competition and also include other fish in their diets. Also, different fish prefer different types of habitats and will move to a new area if foraging conditions in their preferred location decline due to excessive growth of aquatic weeds.

The energy cycle

The energy that supports all life on earth – including life in lakes – originates from sunlight. Vascular plants and phytoplankton (algae) capture light in the chloroplasts of their cells and convert it to energy through photosynthesis. Aquatic plants and phytoplankton use this energy to subsidize new growth, which is consumed and used as an energy source by aquatic fauna. For example, phytoplankton is eaten by zooplankton or vascular plant tissue is eaten by insect larvae. The zooplankton and insect larvae are then eaten by larger insects and/or insect-eating fish. This energy cycle continues with ever-larger organisms consuming smaller ones and provides a vivid illustration of the "trickle-up economics" of energy cycling. As this example demonstrates, the vegetated aquatic habitat that is essential for insects and small fish can be a critical component in the process that fosters growth of harvestable fish.

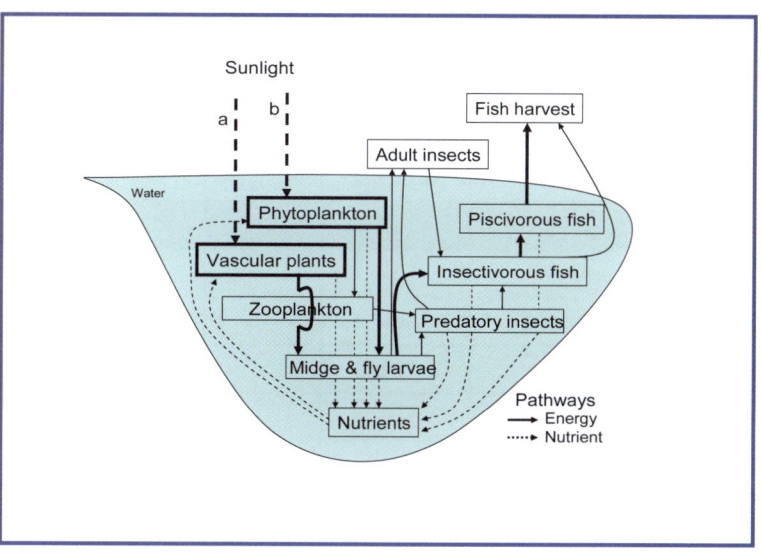

The relationship between fish and aquatic plants

The abundance of some fish declines with increased plant densities. For example, populations of white bass (*Morone chrysops*), gizzard shad (*Dorosomoa cepedianum*) and inland silverside (*Menidia beryllina*) generally decline where heavy vegetation is present. In contrast, many juvenile and some adult fish prefer habitats with aquatic vegetation; in fact, over 120 different species representing 19 fish families have been collected in aquatic plant beds. Sites with vegetation generally have higher numbers of fish compared to non-vegetated areas. In fact, densities of greater than 1 million fish per acre have been reported in areas containing a diversity of aquatic plants. Very few of these fish, however, survive to become large adults, so high numbers of small fish do not always result in populations of large mature fish. Excessive growth of aquatic plants promotes high populations of small fish in contrast to more diverse and balanced plant populations. Reduced plant densities due to weed management activities, boat traffic and/or natural senescence may change or cause the loss of invertebrate food sources. However, studies of lakes where invasive plants were treated with early applications of herbicides to allow native plants to reestablish have revealed that removal of exotic weeds has little impact on invertebrate populations and no measurable effect on fish communities.

Fish and their affinity for plants

Bluegill sunfish (*Lepomis macrochirus*) are often referred to as the "kings" of plant-loving fishes and strongly prefer vegetated habitats throughout much of their lives. There are many different types of

Fish	Plant Affinity	Life stage			Relationship		
		Larvae	Juvenile	Adult	Spawn	Forage	Predator avoidance
Bluegill sunfish	High	X	X	X	X	X	X
Common carp	High	X	X	X	X	X	X
Largemouth bass	High	X	X	X	X	X	X
Musky	High	X	X	X	X	X	X
Northern pike	High	X	X	X	X	X	X
Black crappie	Moderate		X	X	X	X	X
Smallmouth bass	Moderate		X	X		X	X
Yellow perch	Moderate	X	X			X	X
White crappie	Low		X			X	
Salmon, trout	Low		X				X
Shad	Low	X					
Walleye	Low			X		X	

small sunfishes, but the bluegill is likely one of the most popular freshwater fish in North America. The bluegill is the most intensely studied freshwater fish in the US and is considered to be a "lab rat" by fish biologists. In addition to its popularity with scientists, the bluegill has been widely stocked, carefully managed and regularly harvested in natural and artificial systems throughout the US. Bluegill is a premier food fish and is called "pan fish" in the North and "bream" in the South.

Similar to other sunfishes, bluegill often move to new habitats as they age. Bluegill sunfish spawn and nest in colonies near areas of submersed vegetation, where soft sediment and plants are cleared. Bluegill larvae are transparent and can safely move from shallow shoreline habitats to open water where they feed on plankton. As the larvae grow larger and develop color, they become more attractive to predators and seek refuge among aquatic vegetation where they feed on insects, midges and small crustaceans. Juveniles and small adult fish remain among shoreline plants and feed on the food they can capture; as they grow, they may shift to feeding on larger crustaceans, insects and amphipods. As fish mature, grow larger and change color, their chances of being eaten by predators decrease and they shift to more optimal feeding grounds. Bluegill continue to feed in vegetated habitats where they can avoid larger predators until they reach approximately 8" in length. Fish of this size are large enough to escape most of the risk of predation, so these mature bluegill will venture away from the complex structure provided by plants and move to feed in open water. This reduces feeding competition among bluegill and provides access to larger fish that bluegill consume to supply the energy needed for continued growth. Bluegill are not considered herbivores, but they do consume plant material, most likely by accident as they forage for insects and crustaceans living on

aquatic plants. Aquatic plants thus play a critical role in the growth of bluegill sunfish by hosting insects, crustaceans and invertebrates that are eaten by young fish and by providing cover that allows young fish to hide from predators.

Fish populations in lakes with a diverse assemblage of phytoplankton, aquatic plants and habitats tend to be stable. This is a general ecological principle that applies to wildlife, fish and other organisms. However, the bluegill sunfish illustrates why it is unwise to make specific "ironclad" statements regarding the habitat requirements of fish. As noted above, bluegill sunfish have very close associations with aquatic plants but can also become quite large and develop robust populations in managed fish ponds that lack aquatic plants. This apparent conflict is partially explained by the concept that bluegill food webs may be based more on phytoplankton where the predator-prey relationship has been simplified.

Largemouth bass (*Micropterus salmoides*) are stocked throughout the world and are among the world's top freshwater game fishes. Largemouth bass are plant-loving and are closely associated with aquatic plants, spending much of their lives in or around vegetated habitats. Adult largemouth bass diligently protect their nests and offspring from predators. The structure provided by moderate densities of submersed plants improves nesting success, but an overabundance of plants can reduce nesting success. Larvae of largemouth bass feed mostly on microcrustaceans and juveniles consume larger (but still small) crustaceans, whereas mature largemouth bass primarily eat aquatic insects and small fishes (e.g., bluegill, shad and silverside). Aquatic plants serve as critical habitats that support the prey that largemouth bass rely so heavily on through their lives. These prey resources directly or indirectly influence growth and the ability of largemouth bass to overwinter and survive adverse conditions. Therefore, the abundance of largemouth bass is strongly correlated with the abundance of submersed vegetation in its habitat. However, this correlation varies based on the types and densities of the plant species in the habitat.

Smallmouth bass (*Micropterus dolomieu*) prefer deeper, cooler waters with rocks and/or woody cover and generally avoid shallow water that is dominated by aquatic plants. However, like the largemouth bass, young smallmouth bass prey on the insects, crustaceans and other microfauna that are hosted by aquatic plants. More mature smallmouth bass consume crayfish, larger insects and other fishes (including shad). Shad feed primarily on phytoplankton and detritus and avoid aquatic vegetation, so the diet of adult smallmouth and largemouth bass may be dependent on prey fish that do not prefer a vegetated habitat, especially in reservoirs. Smallmouth bass protect their nests and offspring but are less selective of nesting location and will choose nesting sites in shallow water if the water has some form of cover. This cover may be provided by aquatic plants, but most sites have cover in the form of rocky outcrops or overhanging woody debris. Because young smallmouth bass consume microfauna associated with aquatic plants and sometimes use aquatic plants to avoid predators, their relationship with aquatic plants is moderate.

White crappie (*Pomoxis annularis*) have a low affinity for aquatic plants as they typically spawn in nests away from vegetation and spend much of their time as adults and juveniles in open water. However, aquatic plants can directly affect spawning and indirectly influence the diet available to young white crappie. Research suggests that excessive amounts of aquatic plants may reduce spawning success of a nesting colony of white crappie. In addition, the presence of aquatic plants may deter nesting altogether. Eggs of white crappie have been found in aquatic vegetation; however, this is most likely incidental drift of eggs from nearby nesting sites. Larval white crappie feed primarily on microfauna, whereas juveniles feed on insect adults and larvae (i.e., midges and water boatmen) that frequently inhabit vegetated habitats.

Black crappie (*Pomoxis nigromaculatus*) are more closely associated with aquatic plants than their cousins, the white crappie, and have a moderate affinity for plants. Adult black crappie prefer sites with plants – including submersed, emergent, flooded and even inundated terrestrial species – for nesting and spawning and are more likely than white crappie to care for nests and offspring. Like white crappie, they also rely on many of the insects that live in aquatic vegetation. In fact, young black crappie rely heavily on insect larvae and other microfauna that are strongly associated with vegetated habitats.

Gizzard shad (*Dorosoma cepedianum*) are small fish that are widely distributed and are frequently stocked in reservoirs as prey for fish-eating fishes such as crappie and striped, largemouth and other bass. Gizzard shad are not usually considered to be associated with aquatic plants; as larvae, they may rely on food resources from vegetated habitats but their affinity for these habitats is low. Larvae of gizzard shad feed on algae, protozoans and microfauna, whereas adults are more herbivorous and consume phytoplankton in the water column and detritus (decomposed vascular plants) in the sediment. Gizzard shad usually spawn at or near the surface of the water and broadcast their eggs. Eggs drift on the water and can attach to any surface, but it is not uncommon to find egg masses attached to aquatic vegetation. In fact, some egg masses are so large that stems of emergent aquatic plants may collapse under their weight.

Common carp (*Cyprinus carpio*) are invasive, exotic, nuisance species that are detrimental to many aquatic systems. Common carp are frequently found in reservoirs and natural lakes and are associated with shallow areas that have soft sediments and abundant submersed vegetation. Common carp are omnivorous bottom feeders whose diets are composed primarily of organic detritus (mostly in the form of dead plant material) and benthic organisms, including insect adults and larvae, crustaceans, snails, clams and

almost anything else organic that they encounter. The mouth parts of common carp are specialized for foraging for hard items (i.e., plants and animals) within soft sediments and among the roots of aquatic plants. Adult fish typically spawn in shallow water inhabited by aquatic plants, where plant stems and leaves serve as attachment sites for fertilized eggs after spawning. Eggs require oxygen to survive; egg attachment to plant structures prevents eggs from settling into soft sediments that lack the oxygen needed for egg survival.

Salmon and Trout are not usually associated with aquatic plants and their affinity for vegetated habitats is typically thought to be low. However, some trout species may develop indirect relationships to aquatic plant habitats after the fish are introduced into cool reservoirs and natural lakes. For example, the diet of trout in these systems is often dominated by adults, nymphs and larvae of caddisfly, stonefly, cranefly and mayfly, all insects that are frequently associated with aquatic vegetation. This observation, along with reports that navigation and migration of adult salmon and trout may be hindered by dense beds of invasive aquatic plants, suggests that the relationship of salmon and trout to aquatic vegetation may be complex.

Northern Pike (*Esox lucius*). Aquatic plants play an important role in the foraging and reproductive strategies of northern pike, which typically avoid strong currents and have strong affinities for dense beds of aquatic plants during feeding and spawning. Northern pike primarily feed on other fish by using "ambush" foraging strategies—they wait and strike at prey with a burst of swimming energy. Northern pike are among the first fish to spawn in early spring and

broadcast their adhesive-coated eggs on shallow weedy areas. After being released, the eggs drift and settle on submerged vegetation, where they attach and are well-oxygenated.

Muskellunge or Muskie (*Esox masquinongy*) are rarely found far from aquatic plants during any stage of their life. They rely heavily on prey resources (i.e., fish, young ducks, frogs and muskrats) that live in vegetated habitats. Muskie spawn later than northern pike, but utilize similar spawning tactics and rely on plants to successfully reproduce. Eggs of muskie also have an adhesive coating and adhere to plant structures after being broadcast.

Walleye (*Stizostedion vitreum*) are not classified as having a strong affinity for aquatic vegetation, despite reports that walleye are sometimes caught near vegetation. However, vegetation in flooded marshes can provide a substrate for spawning, and populations of some species used by walleye as prey (e.g., yellow perch) do rely on vegetated habitats. Walleye are not tolerant of increases in turbidity or suspended sediment. Therefore, aquatic plants may play an indirect role in improving the walleye habitat in some systems by filtering sediments and decreasing water turbidity.

Adults of **Yellow Perch** (*Perca flavescens*) are typically found in open waters with moderate levels of aquatic plants, but when young their affinity for plants is relatively high. Yellow perch are frequently associated with rooted aquatic vegetation. Successful spawning sites typically contain some form of structure, most often in the form of submerged aquatic plants. Like bluegill, young yellow perch switch habitats as they mature. As clear larvae, they feed in open water on zooplankton; once they become pigmented, they return to shallow water with vegetation where they feed on small fishes and insects along the bottom.

Plants provide critical structure to aquatic habitats

The shade created by leafy plants is important to many visual feeders because shade can improve visibility for both selecting prey and avoiding predators. Vegetated aquatic habitats also provide food for young and small fish of many species while protecting them from predators. The abundance and diversity of aquatic fauna eaten by small fish are higher in vegetated habitats than in areas with no plants because leaves and stems provide a surface for attachment; also, small gaps among plants can provide a place for fauna to escape and hide from predators. As vegetated habitats become more complex, the risk of small fish becoming prey may be decreased. However, the ability of fish to forage declines as vegetated habitats become more complex as well. Visual barriers created by leaves and stems may make it more difficult for fish to find and capture prey, whereas swimming barriers that result from dense vegetation can increase search time by reducing maneuverability and swimming velocity. For example, the rate at which sunfish successfully capture prey declines with an increase in structurally complex vegetated habitats. Some fish have developed tactics to address the negative aspects (i.e., reduced food availability accompanied by increased efforts to capture prey) associated with densely vegetated areas. The largemouth bass, for example, changes foraging tactics in complex habitats and switches from actively pursuing prey to ambushing them as they drift or swim by.

Plants influence growth of fish

Studies have shown that aquatic plant abundance affects the growth and health of fish, especially plant-loving fish such as the sunfishes. Habitats with moderate amounts of aquatic vegetation provide the optimal environment for many fish and enhance fish diversity, feeding, growth and reproduction. Conversely, both limited and excessive plant growth may decrease fish growth rates.

High densities of plants can reduce the growth and health of largemouth bass and of black and white crappie, most likely by reducing foraging efficiency. Fisheries scientists have predicted that largemouth bass growth significantly declines in systems with > 40% coverage of aquatic plants and that maintaining plant beds at an average standing crop of 5 tons of fresh weight per acre (4 ounces per square foot) would improve foraging efficiency of largemouth bass. A total removal of plant biomass exposes forage fish and can, at least temporarily, increase growth of predator fish species (i.e., largemouth bass, black and white crappie, bluegill and other sunfishes) that rely heavily on the prey that inhabits vegetated habitats.

Rapid removal of aquatic plants can alter foraging behaviors and encourage young largemouth bass to switch to eating fish sooner in life, which results in more rapid growth. Conversely, young sunfish grow most quickly in vegetated habitats because when plants are absent or sparse, competition for forage sources increases among these fish; less food resources are available to them and growth slows. However, growth of these fish can also be slowed when plant density is too high, especially in shallow-water areas where plants form monotypic beds.

Plants influence spawning
Studies suggest that the structure provided by plant beds is important to fish reproduction. In fact, many fish in North America are "obligate plant spawners" that directly or indirectly require aquatic plants in order to successfully reproduce. At least a dozen fish families use vegetation as nurseries for their young and reproductive success of nest spawners is improved when they have access to sites with aquatic vegetation and/or some form of structure. Fish can derive a number of benefits from nesting near aquatic plants. For example, vegetation can protect nest sites from wave action and sedimentation that can harm eggs and small fish. Also, parents often use aquatic plant patches or edges as "backing" to protect nests from predators. In addition, many fish that live among aquatic plants are visual feeders and the shade produced by overhanging leaves and plant canopies improves visual acuity so fish can find prey – and avoid becoming prey – with greater success. The shallow areas preferred for spawning by nesting fish are not static and can change over time so that a formerly ideal nesting site can become less than perfect. These areas can become overgrown with aquatic plants, which can hinder optimal spawning. Also, nesting fish can change the composition of the littoral zone by disturbing or altering plant growth, which could affect future nesting success.

Plants influence the physical environment
Aquatic plants can change water temperatures and available oxygen in habitats, thus indirectly influencing growth and survival of fish. The amount of oxygen a fish uses during the course of a day is referred to as daily oxygen consumption rate. High numbers of large fish are not usually found in warm-water habitats that are low in dissolved oxygen because larger fish in warmer water need more oxygen; however, smaller fish are more tolerant of such conditions. Shallow areas where aquatic plants are present and water temperatures increase quickly are inhabited by small fish more frequently than large fish because small fish have lower oxygen consumption rates and can tolerate the reduced oxygen available in these habitats. Dense monotypic beds of weeds in shallow-water habitats can negatively impact fish habitats. The structure resulting from dense growth of stems and leaves can interfere with water circulation and surface exchange of atmospheric oxygen, resulting in high water temperatures and low dissolved oxygen. These conditions can seriously impact fish health; in fact, it is not uncommon to have localized fish kills in areas with extremely dense aquatic weeds. Dense plant beds sometimes have relatively open areas that allow water circulation and oxygen exchange to occur. These areas are usually temporary, but they can serve as important refuges for fish during periods when oxygen levels are low in the rest of the weed bed. Plant beds that are managed for fish habitats should include open areas such as patches and/or lanes to improve the water circulation and oxygen exchange that are important to fish health.

The "perfect" lake: artificial and natural systems

Before determining the optimal amount and type of aquatic plants needed to create "perfect" conditions for fish growth, it is important to recognize that the two types of water systems – artificial and natural – differ from one another and present different challenges for management of aquatic plants. Both types of system can be found throughout the US; as a result, the species of fish that inhabit them (and angler goals) vary by location and contribute to management challenges. As noted above, most fish require some sort of structural habitat at some point in their lives. A diversity of structures provides a diversity of habitats, which can support many different types of aquatic organisms, including numerous species of fish. Therefore, a critical goal in managing artificial and natural water systems should be the maintenance of diverse habitats within the littoral zone, which can be accomplished by ensuring that a variety of plant species are available.

Reservoirs

Reservoirs are typically young (< 100 years old) artificial systems constructed to prevent flooding, generate electrical power and/or to provide navigation for barge traffic. Much of a reservoir is an artificial basin on a flooded – but formerly terrestrial – site; therefore, few reservoirs have naturally occurring populations of native aquatic plants. The sediments of many reservoirs hold seed banks of terrestrial plants that will not germinate under flooded conditions. As a result, the sediment is often a barren benthic mud that provides ideal conditions for invasion by exotic plants. In fact, many reservoirs in the US have been taken oven by aquatic weeds and plant diversity is typically very low.

Fish may naturally inhabit reservoirs, but providing fish habitat is often a byproduct of the reservoir's construction and is rarely intentional. Reservoirs in the southern US are typically stocked with a variety of plant-loving fish, including largemouth bass and bluegill sunfish. As shown earlier, aquatic plants thus play a critical role in the growth of these fish by hosting prey such as insects, crustaceans and invertebrates and by providing cover that allows fish to hide from predators. However, dense monotypic beds of aquatic weeds can restrict the benefits associated with a vegetated habitat by reducing fish foraging ability. This results in a fish population with high numbers of small individuals that fail to grow large, a condition sometimes referred to as a "stunted population." Such populations consist of many individuals feeding in dense habitats which provide better forage resources for smaller individuals, but which restrict foraging opportunities for larger fishes. A plant density that results in coverage of between 20 and 60% of the surface area within the littoral zone generally provides the best fish habitat and recreational opportunities in reservoirs.

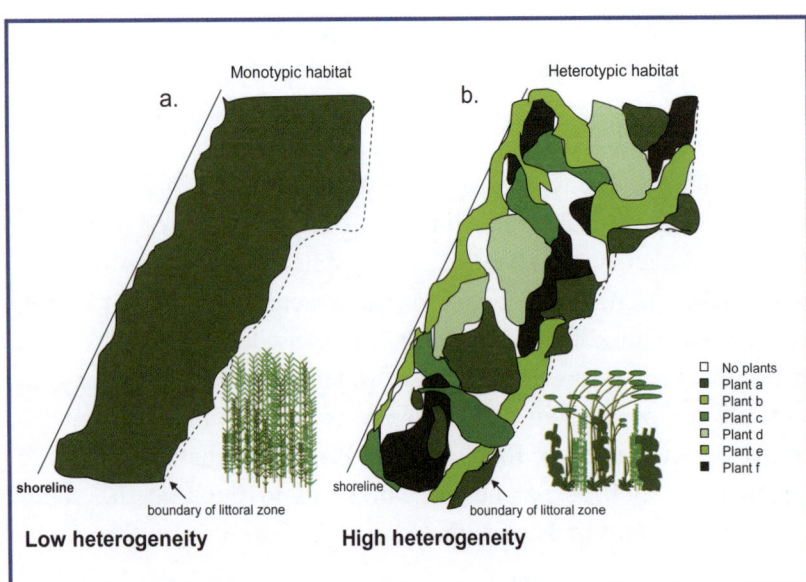

Natural lakes

Many natural lakes form as a result of natural events such as flowing water, earthquakes and animal activities like dam building, but most natural lakes in the northern US are the result of glacial disturbance. These systems were formed many years ago (most recently ten thousand years ago) and are often vegetated by diverse collections of native and endemic aquatic plants. Therefore, management of natural lakes differs significantly from methods used in reservoirs which are usually dominated by monocultures of invasive species.

Natural lakes are diverse in both aquatic plants and fish. Like reservoirs, most of the fish in natural lakes require a structural habitat at some point in their lives. In fact, many are plant-loving fish that choose to spend much of their life feeding and growing in vegetated habitats. The diversity of native and endemic aquatic plants furnishes the littoral zone with a wide variety of structures that differ in size and plant composition, a condition referred to as habitat heterogeneity. This diverse habitat is home to a number of fishes adapted to this environment, including largemouth bass, bluegill, crappie, northern pike, muskie, young perch and walleye.

<u>Summary</u>
Most freshwater fish rely on aquatic plants at some point during their lives and prefer specific habitats based on their growth stage. Young fish use aquatic vegetation as a food source – both by directly consuming plants (in most cases incidentally) and by foraging for the microfauna associated with the plants – and as cover to hide from predators. Mature fish move to more open waters to increase foraging success and consume other fish to supplement their diets. Nesting, growth and foraging success of plant-loving fish are influenced by plant composition and density. While many fish require some aquatic vegetation for optimal growth, excessive amounts of aquatic vegetation can negatively impact growth by reducing foraging success. Also, different fish prefer different types of habitats and will move to a new area if foraging conditions in their preferred location decline due to excessive growth of aquatic weeds.

An "optimal", one-size-fits-all fish habitat is impossible to describe, which leads to confusion and often erroneous conclusions. For example, a crappie fisherman has a different idea of a perfect habitat than does a bass fisherman. The parameters of an ideal habitat change based on the size and species of fish, the type of lake, structures present in the lake and numerous other factors. However, the "optimal" habitat that provides a beneficial environment for most animal populations is one that contains a large diversity of native plants.

<u>For more information:</u>
•Dibble ED, KJ Killgore and SL Harrel. 1996. Assessment of fish-plant interactions. American Fisheries Society Symposium 16:357-372.
•Etnier DA and WC Starnes. 1993. The fishes of Tennessee. The University of Tennessee Press, Knoxville TN.
•Horne A and C Goldman. 1994. Limnology (2nd ed.). McGraw-Hill.
•Jeppesen E, M Sondergaard, M Sondergaard and K Christofferson. 1998. The structuring role of submerged macrophytes in lakes. Springer, New York NY.
•Kovalenko K, ED Dibble and R Fugi. 2009. Fish feeding in changing habitats: effect of invasive macrophyte control and habitat diversity. Ecology of Freshwater Fish 18(2):305-313.
•Lee DS, CR Gilbert, CH Hocutt, RE Jenkins, DE McAllister and JR Stauffer, Jr. 1981. Atlas of North American Freshwater Fishes. Publication #1980-12, North Carolina Biological Survey.
•Mettee MF, PE O'Neil and JM Pierson. 1996. Fishes of Alabama and the Mobile basin. Oxmoor House, Inc., Birmingham AL.
•Ney JJ. 1978. A synoptic review of yellow perch and walleye biology, p1-12, In: RL Kendall, ed. A symposium on selected coolwater fishes of North American. American Fisheries Society Special Publication 11, 437.
•Pflieger WL. 1978. The fishes of Missouri. Missouri Department of Conservation.
•Robinson HW and TM Buchanon. Fishes of Arkansas. 1984. The University of Arkansas Press, Fayetteville AR.

•Ross ST. 2001. The inland fishes of Mississippi. University Press of Mississippi, Oxford MS.
•Scott WB and EJ Crossman. 1973. Freshwater fishes of Canada. Fisheries Breeding of Canada Bulletin 184:1-966.

Photo and illustration credits:
Page 9: Energy cycle; Horne and Goldman 1994
Page 11 upper: Bluegill sunfish; Eric Engbretson, US FWS
Page 11 lower: Largemouth bass; Dean Jackson, professional fisherman
Page 12: Black crappie; Lawrence Page, Florida Museum of Natural History
Page 13 upper: Common carp; Richard A Bejarano, Florida Museum of Natural History
Page 13 lower: Northern pike; Robin West, US FWS
Page 14: Yellow perch; Duane Raver, US FWS
Page 16: Low vs. high heterogeneity; Eric Dibble, Mississippi State University

Ryan M. Wersal
Mississippi State University, Starkville MS
rwersal@gri.msstate.edu

Kurt D. Getsinger
US Army ERDC, Vicksburg MS
Kurt.D.Getsinger@usace.army.mil

Impact of Invasive Aquatic Plants on Waterfowl

Introduction

Studies that evaluate the relationship between waterfowl and aquatic plants (native or nonnative) usually focus on the food habits and feeding ecology of waterfowl. Therefore, the purpose of this chapter is to describe the dynamics of waterfowl feeding in relation to aquatic plants. The habitats used by waterfowl for breeding, wintering and foraging are diverse and change based on the annual life cycle of the waterfowl and seasonal conditions of the habitat. For example, waterfowl require large amounts of protein during migration, nesting and molting and they fulfill this requirement by consuming aquatic invertebrates. A strong relationship exists between high numbers of aquatic invertebrates and diverse aquatic plant communities, so diverse plant communities play an important role in waterfowl health by hosting the invertebrates needed to subsidize

waterfowl migration, nesting and molting. After all, waterfowl native to the US have evolved alongside diverse plant communities that are likewise native to the US and utilize these plants to meet their energy needs. Metabolic energy demands of waterfowl are high during the winter months, so waterfowl need foods that are high in carbohydrates such as plant seeds, tubers and rhizomes during winter. Many ducks will sometimes abandon aquatic plant foraging while on their wintering grounds and feed instead on high-energy agricultural crops such as wheat, corn, rice and soybeans.

The nutritional requirements of waterfowl have historically been met in shallow lakes and wetlands where diverse aquatic plant growth is abundant. It is therefore important to understand the interactions between waterfowl and aquatic plants in order to provide quality habitat throughout migration corridors. The abundance and availability of quality habitat with adequate food, cover and water is the most important ecological component affecting waterfowl populations. In order to support waterfowl health, breeding and survival, the maintenance of quality habitats is crucial so that waterfowl have access to foods they prefer instead of having to feed on what is available.

The preferred food habitats and feeding ecology of waterfowl differ based on the group of waterfowl (i.e., dabbling ducks, diving ducks, or geese and swans). For example, dabbling ducks (also called puddle ducks) vary greatly in size and "tip up" during feeding. Their feeding is constrained by how far their necks can reach into the water column (12 to 18") and depth of the water, so dabbling ducks prefer habitats with shallow water and/or moist soil. Diving ducks typically dive (as their name implies) to feed on benthic organisms such as clams and snails or to forage in sediments for tubers and rhizomes of aquatic plants. Geese and swans are the largest of the waterfowl and typically consume more plant material than dabbling ducks and divers; however, as the availability of natural habitats is diminished, geese and swans have shifted from primarily feeding in wetlands to extensive grazing in agricultural areas.

Dabbling (puddle) ducks

Dabbling waterfowl include such species as the mallard (*Anas platyrhynchos*), blue-winged teal (*Anas discors*), green-winged teal (*Anas crecca*), wood duck (*Aix sponsa*), gadwall (*Anas strepera*), northern pintail (*Anas acuta*), northern shoveler (*Anas clypeata*) and American widgeon (*Anas americana*). Most dabbling species are non-selective in their feeding habits and feed primarily on aquatic or moist-soil vegetation that is abundant in a given location. Dabblers will alter their diets as necessary to take advantage of food resources that are available and abundant. Food selection by dabbling ducks often changes based on the season and energy requirements of the waterfowl. Protein is important during spring and summer to ensure breeding success, so invertebrates are critical components in the diet of dabbling waterfowl during these seasons. In late fall and winter, dabblers consume plant material that is high in carbohydrates so they can maintain energy levels and generate body heat throughout the winter months. Dabbling waterfowl utilize submersed plant species as carbohydrate sources to fulfill their energetic demand. Most consume seeds as their primary food source, but some species (mainly widgeon and gadwall) use vegetative parts of plants as well. Also, the specialized bill structure of the shoveler, or spoonbill, allows for sifting and consumption of planktonic algae, which are high in carbohydrates.

Submersed plant communities play important roles in the annual life cycle of dabbling waterfowl. These communities are a direct source of food and also serve as an environment that supports a diversity of aquatic invertebrates. The primary submersed aquatic plants consumed by dabblers are the native pondweeds (*Potamogeton* spp. and *Stuckenia* spp.). The fruits, seeds, starchy rhizomes and winter buds of these species are favored carbohydrate sources for dabbling waterfowl, and sago pondweed (*Stuckenia pectinata*) is reportedly one of the food plants most sought after by these waterfowl. Sago pondweed is likely the single most important waterfowl food plant in the US and often accounts for a significant proportion of the food consumed by fall staging waterfowl, pre-molting waterfowl, flightless molting waterfowl and ducklings.

Diverse plant communities with a wide variety of submersed, floating and emergent plants have more architectural structure and habitat for invertebrates, which results in a greater selection of food sources for dabbling waterfowl. Water bodies that are infested with nonnative species such as hydrilla (Chapter 13.1), Eurasian watermilfoil (Chapter 13.2) and curlyleaf pondweed (Chapter 13.7) lack the habitat complexity required to support diverse invertebrate communities and are not preferred feeding areas for dabbling waterfowl. These nonnative species form dense canopies at the surface of the water, reduce native plant diversity and reduce the carrying capacity of the ecosystem. Also, if large portions of the littoral zones of several water bodies within an area are infested with nonnative plants, waterfowl may be required to continually move in search of adequate forage and resting areas. This constant movement results in poor body condition since high expenditures of energy impact wintering, migration and/or breeding fitness. Birds that are in poor body condition when returning to northern breeding grounds may have reduced nesting success or may not nest at all. Some dabbling ducks such as the wood duck nest in tree cavities, whereas other dabbling waterfowl nest in upland prairie habitat, so nonnative emergent plant species such as purple loosestrife (Chapter 13.6) and phragmites (Chapter 13.9) would not impact nest site selection for dabblers as it does for some diving species of waterfowl. However, if shallow wetlands and moist-soil areas become infested with invasive emergent weeds, the quality of food and refuge habitat for ducklings and molting waterfowl could be diminished during summer months and could ultimately reduce survival. For example, ducklings and smaller species of dabbling waterfowl such as blue and green-winged teal feed in moist soil and in areas where water depths do not exceed 8 to 12 inches. As a result, dense infestations or monotypic stands of invasive weeds can limit foraging efficiency and food quality for these ducks.

Diving ducks

Common diving ducks in North America include canvasback (*Aythya valisineria*), redhead (*Aythya americana*), lesser scaup (*Aythya affinis*), greater scaup (*Aythya marila*), ring-necked duck (*Aythya collaris*), bufflehead (*Bucephala albeola*) and common goldeneye (*Bucephala clangula*). Sea ducks and mergansers will not be discussed because sea ducks are rarely observed on inland waters and mergansers mainly consume fish.

The diet structure of diving ducks is similar to that of dabbling waterfowl because diving ducks also rely on aquatic plants, their diet alternates with the annual life cycle of the birds and food selection is influenced by gender. Female diving ducks typically consume more invertebrates during nesting, incubation and brood rearing to maintain the protein and fat stores that result in good body condition. In contrast, male diving ducks (particularly older juvenile and adults) tend to consume more plant material. Canvasback ducks feed primarily on seeds and tubers of pondweeds and the native submersed plant vallisneria (*Vallisneria americana*), from which the bird takes part of its Latin name.

Vallisneria is widely distributed and is considered the most important food source for canvasback ducks. Displacement of native vallisneria by invasive plants such as Eurasian watermilfoil or hydrilla will impact canvasback foraging behavior and can lead to annual fluctuations in canvasback populations. Canvasback numbers could decline or expand depending on the quality and abundance of vallisneria-dominated communities, which is linked to competition with invasive plants. Pondweeds are also very important food sources for redhead and ring-necked ducks, but these two species forage in shallow-water areas more frequently than other types of diving waterfowl and therefore consume a diversity of plant material. Ring-necked ducks feed heavily on wild rice (*Zizania palustris*), coontail (*Ceratophyllum demersum*), sedges (*Carex* spp.), rushes (*Scirpus* spp.) and the seeds and tender submersed shoots of the floating plant watershield (*Brasenia schreberi*). However, divers such as ring-necks are highly adaptive foragers and will reportedly feed on hydrilla tubers if hydrilla populations are abundant on their wintering grounds, particularly in large inland water bodies in Florida. The two species of scaup generally consume more invertebrates than plant matter, but plants do become important to scaup during fall and winter. With the exception of the ring-necked duck, all diving waterfowl will readily switch to feeding on mussels and clams in southern wintering grounds if plant material is limited.

Nonnative submersed weeds such as hydrilla, Eurasian watermilfoil and curlyleaf pondweed would also have an impact on feeding activities of diving waterfowl. Since native pondweeds comprise a considerable portion of the food consumed by diving waterfowl, any reduction in the abundance or richness of these native plant species would have an adverse impact on waterfowl in that area. Diving waterfowl will reportedly consume the seeds of Eurasian watermilfoil and tubers of hydrilla; however, these observations were reported in areas heavily infested with these weeds and waterfowl were forced to forage on dense stands of these exotic plants, as their preferred native species were unavailable. It should also be noted that some propagules such as seeds can pass through the digestive tract of waterfowl and still be viable. Even if waterfowl utilize nonnative plants as food sources, this may result in long-distance dispersal and spread of aquatic weeds to other areas of the country. Water bodies should be managed to promote the growth of a diversity of native aquatic plants because these are most utilized by diving waterfowl and they provide habitat for greater numbers and species of invertebrates.

Diving species of waterfowl also require emergent aquatic plants for nesting habitat. Canvasbacks and redheads nest almost exclusively above the water in specific types of vegetation. Hardstem bulrush (*Scirpus acutus*), cattails (*Typha* spp.), bur-reed (*Sparganium* spp.) and sedges that extend 1 to 3 feet above the water surface are preferred habitat for nesting. These plant species generally have more succulent and flexible stems that waterfowl can manipulate for nest construction. Nonnative plant species such as purple loosestrife and phragmites have hardened, woody stems that do not support waterfowl nesting. Purple loosestrife and phragmites will also outcompete native plants preferred for nesting, which further reduces breeding habitat that is becoming scarce due to pressure from human development and agricultural practices.

Geese and swans

Geese (Canada, snow and white-fronted) are primarily vegetarian and have shifted their feeding ecology toward agricultural grains and/or green-fields, including golf courses and parks. For

example, corn and wheat have provided the majority of food for migrating and wintering Canada geese in recent decades and rice is frequently consumed by geese in the southern US. When agricultural grains become scarce in late winter, geese will feed on the green tissue of native moist-soil plants such as millets (*Echinochloa* spp.), smartweeds (*Polygonum* spp.), cut-grasses (*Leersia* spp.) and spike rushes (*Eleocharis* spp.). This switch in food sources also corresponds to times when crude protein is needed for migration and nesting. Swans are also primarily vegetarian but feed on aquatic plants more than do geese. The diets of swans are based primarily on wigeongrass (*Ruppia maritima*), pondweeds and vallisneria during the winter months, but swans will forage in agricultural fields, golf courses or urban lawns when populations of aquatic plants are depleted.

Summary

Dabbling ducks, diving ducks, geese and swans are generalists and will consume the food sources available in a given area. Waterfowl prefer to forage and rest in shallow-water habitats that support diverse communities of submersed plants, including nonnative species. However, waterfowl usually prefer native species of aquatic and moist-soil plants to nonnative, invasive vegetation. Dabbling waterfowl prefer seeds of smartweed, millet, pondweeds, sedges and rushes, as well as invertebrates that typically thrive in association with these plants. Although waterfowl will utilize nonnative plants, these species are generally not preferred and are consumed only because they are locally abundant. Diving ducks such as canvasbacks and redheads rely heavily on pondweeds and vallisneria, but nonnative aquatic weeds such as Eurasian watermilfoil, hydrilla and curlyleaf pondweed can outcompete and reduce the presence of these valuable and desirable native plants. Furthermore, dense infestations of nonnative emergent species such as purple loosestrife and phragmites reduce the already-dwindling nesting habitat for many waterfowl species. North American waterfowl have evolved and thrive in habitats that support a variety of diverse native aquatic plants and management should focus on removing monotypic stands of nonnative plants to promote native plant growth.

For more information:

- Baldassarre GA and EG Bolen. 1994. Waterfowl ecology and management. John Wiley and Sons, New York.
- Havera SP. 1999. Waterfowl of Illinois: status and management. Illinois Natural History Survey Special Publication 21.
- Johnson FA and F Montalbano. 1984. Selection of plant communities by wintering waterfowl on Lake Okeechobee, Florida. Journal of Wildlife Management 48(1):174-178.
- Johnson FA and F Montalbano. 1987. Considering water habitat in hydrilla control programs. Wildlife Society Bulletin 15:466-469.
- Kantrud HA. 1986. Effects of vegetation manipulation on breeding waterfowl in prairie wetlands, a literature review. US Fish and Wildlife Service Technical Report 3.
- Kantrud HA. 1990. Sago pondweed (*Potamogeton pectinatus* L.): a literature review. US Fish and Wildlife Service Resource Publication 176.
- Martin AC and FM Uhler. 1939. Food of game ducks in the United States and Canada. US Department of Agriculture Technical Bulletin 634.
- McAtee WL. 1918. Food habits of the mallard ducks of the United States. US Department of Agriculture Technical Bulletin 720.
- Perry MC and FM Uhler. 1982. Food habits and distribution of wintering canvasbacks on Chesapeake Bay. US Fish and Wildlife Service, Patuxent Wildlife Research Center.
- Smith LM, RL Pederson and RM Kaminski (eds). 1989. Habitat management for migrating and wintering waterfowl in North America. Texas Tech University Press, Lubbock TX.
- Soons MB, C van der Vlugt, B van Lith, GW Heil and M Klaassen. 2008. Small seed size increases the potential for dispersal of wetland plants by ducks. Journal of Ecology 96:619-627.

Photo and illustration credits:

Page 19: Pintail drake; George Gentry, US FWS
Page 21: Lesser scaup; Dave Menke, US FWS
Page 22: Snow geese; Lee Karney, US FWS

Chapter 3

Chapter 4

Mark V. Hoyer
University of Florida, Gainesville FL
mvhoyer@ufl.edu

Impact of Invasive Aquatic Plants on Aquatic Birds

Introduction

Birds that live at least part of their lives in or around water are referred to as aquatic birds and/or water birds. Each species has specific requirements that must be met in order to reproduce, survive, grow and reproduce again. It can be challenging to make broad statements that apply to all aquatic birds, but they are often grouped into subclasses based on habitat preference, which allows generalizations to be made about birds with similar requirements. Waterfowl are discussed in Chapter 3, but other groups of aquatic birds that use similar habitats include marsh birds, shorebirds and wading birds.

Marsh birds live in or around marshes (treeless wet tracks of grass, sedges, cattails and other herbaceous wetland plants) and swamps (wet, soft, low, water-saturated land that is dominated by trees and shrubs). This is a broad category that includes many unrelated species of birds, all of which prefer to nest and/or live in marshy, swampy areas. Marsh birds include herons, storks, ibises, flamingoes, cranes, limpkins and rails.

Shorebirds inhabit open areas of beaches, grasslands, wetlands and tundra. These birds, which include plovers, oystercatchers, avocets, stilts and sandpipers, are often dully colored and have long bills, legs and toes.

Wading birds generally do not swim or dive for prey, but instead wade in shallow water to forage for food that is not available on shore. Wading birds include herons, egrets, spoonbills, ibises, cranes, stilts, avocets, curlews and godwits. These birds generally have long legs, long bills and short tails, which allows them to strike and/or probe under the water for fish, frogs, aquatic insects, crustaceans and other aquatic fauna.

It is easy to see that some birds can fall into several of these general groups, so care should be taken when interpreting statements applied to birds in these groups. These subclasses group birds based on habitat preference, but birds are complex, adaptable animals. Thus, regardless of habitat, it may be possible to observe many different aquatic bird species if adequate food sources are available. The purpose of this chapter is to describe how aquatic birds are related to lake morphology, water chemistry and aquatic plants in lake systems and how the presence of large monocultures of exotic invasive plants such as hydrilla (Chapter 13.1) and phragmites (Chapter 13.9) may impact aquatic bird communities.

Lakes and aquatic bird communities

Birds are an integral part of all lake systems, but their role in the ecology of lakes has frequently been overlooked. This is surprising, since aquatic birds are often the first wildlife that is seen when visiting a lake and the vast majority of people who visit lakes enjoy their beauty and grace. However, the majority of earlier research and management conducted on lake systems involved nutrient enrichment problems and aquatic plant management. The focus of this early research was primarily to provide potable water, flood control, navigation, recreational boating, swimming and fishing and consideration was seldom given to aquatic bird communities that utilized these lakes. As a result, little information is available regarding how these different lake management activities affect aquatic bird communities.

This situation began to change rapidly in the 1980s when many ornithologists (scientists studying birds) and limnologists (scientists studying freshwater systems) became increasingly conscious of the importance of birds to aquatic systems. These researchers have worked together to identify many significant relationships between lake limnology and aquatic bird populations. This research can be used to predict the impact of habitat changes resulting from invasion by aquatic weeds and from lake management programs on aquatic bird communities.

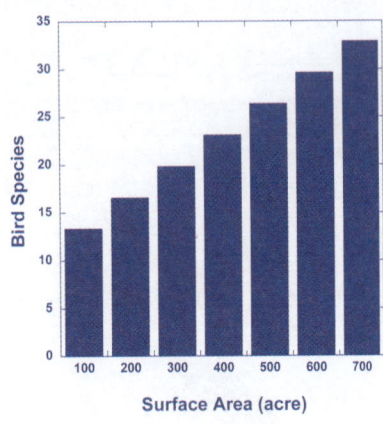

Lake area and aquatic bird species richness

There is a strong relationship between bird species richness (the number of bird species in an aquatic community) and the surface area of the lake they inhabit. Many studies have shown that plant and animal species richness increases as habitat area increases. Most researchers and lake managers agree that larger areas are more likely to include diverse habitats that allow more species niches. Based on this theory, the invasion of a lake system by an exotic species and the resulting monoculture of a single aquatic plant would decrease other environmental niches and would decrease the number of species of aquatic plants and ultimately aquatic birds using that lake system. However, there are few studies that document this type of impact of aquatic weeds on bird populations.

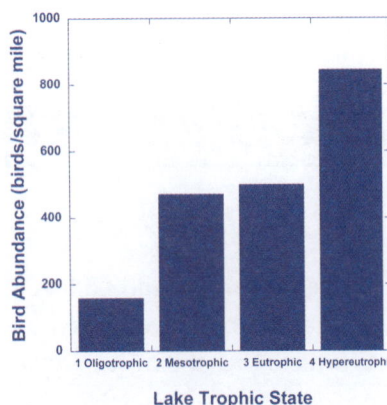

Lake trophic state and aquatic bird abundance

Lake trophic state is the degree of biological productivity of a water body. Biological productivity generally describes the amount of algae, aquatic plants, fish and wildlife a water body can produce. The level of trophic state is usually set by the background nutrient concentrations of the geology in which the lake lies, because nutrients (primarily phosphorus and nitrogen) are the most common factors limiting growth of algae and plants that form the base of the biological food chain (Chapter 1). It is therefore not surprising that lakes with higher trophic states generally support more aquatic birds, since these lakes usually have an abundance of plants and animals that can be used for food and shelter by aquatic birds. Some question whether aquatic birds show up because a lake is productive or whether the lake becomes productive because birds bring nutrients to the system. There have been instances where large flocks of birds such as geese feed on terrestrial agricultural grains and then roost on a lake, ultimately causing elevated nutrient concentrations in a lake. However, most current research suggests that the majority of aquatic bird communities extract their nutrients from the lake and function more as nutrient recyclers than as nutrient contributors.

Most lake management efforts are directed toward the manipulation of lake trophic state, with most resources focused on reducing nutrients caused by anthropogenic activities. However, management agencies in some areas will actually add fertilizer (nutrients) in an attempt to increase productivity of plants, algae and fish, which increases angling activities. In either case, changes to the trophic state of a lake system will have a corresponding impact on the aquatic birds that utilize the lake. If aquatic birds are an important component of an individual lake, this relationship needs to be considered before nutrient manipulations occur.

Aquatic plants and aquatic bird communities

Aquatic birds rely on aquatic plants to meet a large variety of needs during their life cycles. Some birds nest directly in aquatic plants, whereas others use plants as nesting material, foraging platforms, for resting and for refuge from predators. Aquatic plants are eaten by some bird species; in addition, some plants support attached invertebrates that are used as a food source by some aquatic birds. Since there are so many associations between the needs of aquatic birds and aquatic plants, it would be reasonable to expect a strong relationship between the abundance of all aquatic birds and the abundance of aquatic plants in a lake system. However, multiple studies have found no such relationship after accounting for differences in lake trophic state. This surprising lack of relationship between total bird abundance and total abundance of aquatic plants can be explained by the fact that individual bird species require different types and quantities of aquatic plants. Research has suggested that aquatic bird species can be divided into three general groups: 1) birds that are directly related to the abundance of aquatic plants, 2) birds that are negatively affected by an abundance of aquatic plants, and 3) birds that have no relationship to the

total abundance of aquatic plants but require the presence of a particular plant type for completion of their life cycle. However, these are loose generalizations and individual species of aquatic birds can transcend these plant groupings depending on the given lake system and the bird's life requirements.

Birds that are directly related to the abundance of aquatic plants.
Many waterfowl, including the coots and ring-necked ducks described in Chapter 3, use aquatic plants as a food source and thus are generally more abundant in lakes with an abundance of aquatic plants. Other aquatic birds that prefer a habitat with plentiful aquatic plants include limpkins and curlews. These species are generalized feeders that consume insects, fish, small animals, snails and other aquatic fauna that are associated with aquatic vegetation. Limpkins and curlews are often observed walking on and foraging in floating

aquatic plants, waterhyacinth (Chapter 13.5), salvinia (Chapter 13.4), native waterlilies and other plants when this vegetation is present in densities sufficient to support the weight of the birds. If this type of habitat is not available, these birds will forage along sparsely vegetated shorelines and mudflats where water is shallow enough to allow wading. Birds in this group prefer lakes with an abundance of aquatic plants; however, these species will often locate and feed in more diverse habitats when their preferred environment is not available to them.

Birds that are negatively affected by an abundance of aquatic plants. Some bird species, such as snakebirds (*Anhinga anhinga*) and double-crested cormorants (*Phalacrocorax auritus*), must swim through the water to catch fish, crayfish, frogs and other aquatic fauna. Large amounts of aquatic vegetation interfere with the feeding ability of these aquatic birds; therefore, these types of birds tend to decrease in abundance when submersed weeds become abundant in a lake system. Other aquatic birds that prefer sparsely vegetated water are the threatened piping plover (*Charadrius melodus*) and the endangered interior least tern (*Sterna antillarum athalassos*). These species once fed, nested and were abundant on sandbars along the Missouri and Platte Rivers and in other similar areas in the central and northern US; however, piping plovers and interior least terns have experienced major population declines in the last 50 years. Dredging and damming of rivers has destroyed most of the sandbar habitat preferred by these species and flood control projects have reduced scouring and re-forming of new sandbars. In addition, old sandbars have become densely vegetated, further reducing the nesting and feeding grounds required by these aquatic birds. This is particularly problematic in the Midwest, where phragmites and purple loosestrife (Chapter 13.6) have invaded most sandbars formerly inhabited by piping plovers and interior least terns.

Some aquatic birds are only affected by certain types of aquatic weeds. For example, eagles and ospreys soar over open water in search of fish swimming near the surface of the lake, so submersed aquatic weeds rarely hinder feeding by these species. In fact, since submersed plants reduce wind and wave action and improve water clarity, the presence of these aquatic plants may actually increase the feeding efficiency of sight feeders such as eagles and ospreys. However, dense populations of floating plants and floating-leaved plants (e.g., waterhyacinth, salvinia, waterlilies, etc.) may negatively impact the foraging success of sight feeding aquatic birds because fish are hidden beneath the vegetation. Sight feeders may be forced to abandon lakes that are heavily vegetated with these types of plants and seek out new habitats with open water that provide an unobstructed view of their prey.

Birds that have no relationship to the total abundance of aquatic plants but require the presence of a particular plant type for completion of their life cycle. Some aquatic bird species – including the secretive American bittern (*Botaurus lentiginosus*) and least bittern (*Ixobrychus exilis*) – require tall, emergent vegetation like cattails and bulrush for concealment from predators regardless of the total amount of aquatic vegetation present in the lake. Both species of bittern "freeze", with neck outstretched and bill pointed skyward, when danger threatens and sway in imitation of wind-blown emergent vegetation such as cattails. Even nestling least bitterns, still covered with down, adopt this posture when threatened. Invasion by exotic species of aquatic plants would probably not impact this type of bird species unless the exotic species reduces the abundance of the required aquatic plant.

Many wading birds also fall into this group and do well in lakes regardless of the amount of aquatic plants, but one factor that may limit the success of these wading birds is the availability of water shallow enough for them to forage for food. Wading birds that inhabit lakes regardless of the abundance of aquatic plants include great blue heron (*Ardea herodias*), great egret (*Ardea alba*), snowy egret (*Egretta thula*), little blue heron (*Egretta caerulea*) and tricolored heron (*Egretta tricolor*). Larger wading birds can forage in water of greater depths, which increases the area available for foraging. Therefore, the great blue heron has an advantage over the smaller little blue heron in open water. However, larger wading birds may become tangled in vegetation when an invasive exotic species covers a lake; on the other hand, many of the smaller wading birds can actually wade on top of dense plant growth, which vastly increases their foraging area.

Summary

Aquatic birds come in an almost infinite number of sizes and shapes and require many different resources to complete their life cycles. A number of generalizations can be made regarding groups of similar bird types, but it is important to remember that all species are somewhat different. Also, individual species are adaptable and often able to use available resources even if those resources are not preferred. Encroaching invasive exotic plants can increase, decrease or have little impact on a particular aquatic bird, which makes it difficult to predict the impact of aquatic plants on a given species. This dilemma becomes even more challenging when you consider that birds fly and can easily travel from lake to lake to find the habitat that best suits their needs, even though the distance seems prohibitive.

For more information:

•Ehrlich PR, DS Dobkin and D Wheye. 1988. The birders handbook. A field guide to the natural history of North American birds. Simon and Schuster Inc. New York.
•Hanson AR and JJ Kerekes. 2006. Limnology and aquatic birds. Proceedings of the fourth conference working group on aquatic birds of Societas Internationalis Limnologiae (SIL). Springer, Dordrecht, The Netherlands.
•Kerekes JJ and B Pollard (eds.). 1994. Symposium proceedings: aquatic birds in the trophic web of lakes. Sackville, New Brunswick, Canada. Aug. 19-22, 1991. Developments in hydrobiology, vol 96. Reprinted from Hydrobiologia, vol. 279/280.
•Peterson RT. 1980. A field guide to the birds east of the Rockies. Fourth edition. Houghton Mifflin Company. Boston.
•Terres JK. 1980. The Audubon Society encyclopedia of North American birds. Alfred A. Knopf. New York.

Photo and illustration credits:

Page 25: Tricolor heron; Mark Hoyer, University of Florida Fisheries and Aquatic Sciences
Page 26: Graphs; Mark Hoyer, University of Florida Fisheries and Aquatic Sciences
Page 27: Red winged blackbird nest; Mark Hoyer, University of Florida Fisheries and Aquatic Sciences
Page 29: Least bittern; Mark Hoyer, University of Florida Fisheries and Aquatic Sciences

Chapter 5

James P. Cuda
University of Florida, Gainesville, FL
jcuda@ufl.edu

Aquatic Plants, Mosquitoes and Public Health

Introduction

Approximately 200 species of aquatic plants are classified as weeds in North America and nearly 50, or 25%, are considered to be of major importance. Aquatic plants become weedy or invasive when they exhibit rapid growth and produce dense monocultures that displace more desirable native plants, reduce biodiversity, interfere with flood control, impede navigation and create breeding sites for disease-vectoring mosquitoes.

Mosquitoes are insects that belong to the family Culicidae in the order Diptera, or true flies. They are similar in appearance to other flies except they have fragile bodies and their immature stages (eggs, larvae and pupae) develop entirely in aquatic environments. These insects are serious pests that have plagued civilizations throughout human history. In addition to their annoying and often painful bites, they transmit some of the world's most devastating diseases – dengue, encephalitis, yellow fever, dog heartworm and the dreaded malaria. According to a recent report from the University of Florida, more than 500 million new cases of malaria are reported worldwide each year, resulting in about 1 million deaths. Most of the deaths that are caused by malaria are in children under 10 years of age. The importance of mosquitoes from a nuisance and public health perspective cannot be overstated.

Malaria

Malaria was endemic in the US until around 1950 when window screens, air conditioning and mosquito control efforts essentially eliminated malaria in this country. Malaria is caused by four species of a protozoan parasite in the genus *Plasmodium*. This parasite, which is transmitted by a mosquito bite, destroys red blood cells and causes fever, chills, sweating and headaches in infected humans. If not treated, individuals that have become infected with malaria may go into shock, experience kidney failure and eventually slip into a coma and die. The disease is transmitted by several species of *Anopheles* mosquitoes, which are permanent water mosquitoes (see below). These species are widespread and are most abundant from early spring (April) to early fall (September). Until recently, reported cases of malaria in the US were from travelers and returning military personnel who contracted the disease outside the country. However, cases of malaria occur periodically in the US when indigenous *Anopheles* mosquitoes transmit the disease from an infected human who traveled abroad to an uninfected human.

Dengue fever

Dengue is a viral disease, often referred to as "breakbone fever". Symptoms of this mosquito-transmitted disease include headaches, high fever, rash, backache and severe pain in the joints. The excruciating joint pain gives rise to the common name. Disease symptoms usually occur about a week after a susceptible human has been bitten by an infected mosquito and rarely result in death. However, because four strains of dengue virus are recognized, exposure of a previously infected individual to a different strain of dengue virus may result in a more severe case of dengue known as dengue hemorrhagic fever (DHF). There has been an increase in the incidence of DHF in the Western Hemisphere during the last 20 years, with outbreaks occurring in the Caribbean region. Ideal conditions for dengue transmission are present in the southern US. The virus often is "imported" by people entering the country from the tropics. Also, the potential mosquito vectors (yellow fever mosquito, *Aedes aegypti,* and the Asian tiger mosquito, *Aedea albopictus*) are commonly found in close association with humans, breeding in natural and artificial water-holding containers near homes and businesses.

Encephalitis

Encephalitis means inflammation of the brain and is a disease of the central nervous system. Although there are several possible causes for encephalitis, one of the most important involves mosquitoes. Mosquito-transmitted viruses are commonly referred to as arthropod-borne or arboviruses. There are six major types of arboviral encephalitis in the US: California encephalitis (CE), Eastern equine encephalitis (EEE), St. Louis encephalitis (SLE), Venezuelan equine encephalitis (VEE), Western equine encephalitis (WEE) and West Nile virus. These viruses are normally diseases of birds or small mammals and each is caused by a different virus or virus complex. Humans and horses are considered "dead end" hosts for these viruses as there is little chance of subsequent disease transmission back to mosquitoes. However, human and horse cases of arboviral encephalitis range from mild to severe, with permanent damage to the central nervous system or even death. Mosquito genera involved in the transmission of arboviruses include *Aedes, Anopheles, Culex, Culiseta, Ochlerotatus, Coquillettidia* and *Psorophora.*

Yellow fever

Like dengue fever, the yellow fever virus is transmitted primarily in urban areas by the container-breeding mosquitoes *Aedes aegypti* and *Aedes albopictus.* But unlike dengue, the effects on humans are more severe. During outbreaks, the human fatality rate often exceeds 50% of the affected population. Fortunately, the yellow fever virus is restricted to parts of Africa and South America. The likelihood of the yellow fever virus causing an epidemic in the US is extremely low for several reasons. First of all, yellow fever is a quarantinable disease; the Centers for Disease Control and Prevention in Atlanta continually monitor disease outbreaks in the Western hemisphere. Secondly, travelers planning to visit parts of Africa and South America where the virus is endemic are vaccinated to prevent infection. Finally, humans moving to virus-free areas from locations where the virus occurs naturally are required to be vaccinated to prevent transmission.

Heartworms

The filarial nematode (microscopic worm) *Dirofilaria immitis* is responsible for dog heartworm, a serious mosquito-transmitted disease that affects all breeds of dogs. Although the disease occurs in temperate regions of the US, it is more of a concern along the Atlantic and Gulf Coasts from Massachusetts to Texas. If left untreated, the infection rate in dogs can range from 80 to 100%. Foxes and coyotes probably serve as reservoirs for the disease. Cats and humans also can be infected but the parasite is unable to complete its development in humans. Mosquitoes in most of the common genera, including *Aedes, Anopheles, Culex, Ochlerotatus, Mansonia* and *Psorophora*, are capable of transmitting the disease. The life cycle of dog heartworm begins when an infected mosquito feeds on a dog. Juvenile worms (microfilariae) emerge from the mouthparts of the feeding mosquito and enter the dog's skin. The worms migrate in the muscle tissue for 3 to 4 months, penetrating blood vessels and eventually making their way to the right ventricle of the dog's heart, hence the name "dog heartworm". The worms reach maturity in around 5 months; adult female worms measure about 1 foot in length whereas males are only 6 inches long. The life cycle is completed when the adult female produces microfilariae that circulate in the blood and are ingested by a mosquito during a blood meal. Medication for preventing dog heartworm is available from veterinarians.

The role of aquatic plants in mosquito outbreaks

The aquatic stages of most mosquitoes are *not* adapted to life in moving waters. They require quiet pools and protected areas where they can obtain oxygen at the water surface via a single air tube (or siphon) in the larval stage or two tubes (or horns) in the pupal stage. Aquatic weed infestations create ideal habitats for mosquito development because the extensive mats produced by many weeds reduce the rippling effect of the water surface. Some mosquito species even have a modified air tube that they insert into the roots of aquatic plants to obtain oxygen. This protects them from light oils that are applied to the water surface for mosquito control.

From a mosquito control perspective, there are two major larval habitat categories that are of concern to aquatic plant managers: standing water (permanent and temporary) and flood water (detention and retention areas). Permanent water mosquitoes (e.g., species in the genera *Anopheles, Culex, Coquillettidia* and *Mansonia*) are associated with aquatic plants in freshwater marshes, lakes, ponds, springs and swamps. Temporary water mosquitoes (e.g., species in the genera *Culiseta, Ochlerotatus* [=*Aedes*] and *Psorophora*) are associated with vegetation in saline or brackish ditches, borrow pits and canals and freshwater drainage ditches which alternate between wet and dry based on water use and rainfall events.

Permanent water

The amount and type of vegetation occurring in a permanent water body is a good indicator of its potential to produce mosquitoes. For example, the presence of floating mats of cattails, torpedograss, alligatorweed or para grass suggest that larvae of permanent water mosquitoes are likely to be present. Also, dense stands of aquatic plants create ideal conditions for mosquito development by restricting water flow in drainage and irrigation ditches.

Flood water

Detention and retention systems are artificial ponds designed to capture flood water from rainstorm events and filter it before it enters natural systems. Construction of storm water detention/retention areas has increased dramatically throughout the US and they are often required by law for all new commercial and residential developments. Detention ponds differ from retention ponds by the length of time they are "wet." Detention ponds dry out only during drought conditions, whereas retention ponds are designed to dry out rapidly, usually within 72 hours. Under the right conditions, both types of flood control systems can produce aquatic vegetation that can foster mosquito outbreaks. Unless they are properly managed, detention/retention areas overgrown with aquatic vegetation can lead to serious mosquito problems. Detention ponds normally do not produce many mosquitoes unless they alternate between the wet and dry cycles that are required to produce floodwater mosquitoes.

However, if they are not properly managed, they often are invaded by floating and rooted aquatic plants. The only way to prevent a mosquito problem in residential and commercial detention/retention areas that contain these mosquito-producing plants is to control the plants.

Mosquitoes associated with specific aquatic plants

Some species of mosquitoes are associated with certain species of aquatic plants. For instance, the permanent water mosquito species *Coquillettidia pertubans, Mansonia dyari* and *M. titillans* are always associated with waterlettuce, waterhyacinth (Chapter 13.5) and cattails. The extensive fleshy root systems of these species provide an ideal substrate for *Mansonia* larvae to attach and obtain oxygen through air tubes they insert into the plant roots. Also, the fleshy root system of cattail often harbors larvae of *Coquillettidia* mosquitoes. The roots of cattails and other plants also afford mosquito larvae some measure of protection from predators (including fish), as they are hidden from them. Other plants are good indicators of areas likely to produce floodwater mosquitoes. For example, sites with grasses, sedges and rushes often host enormous numbers of *Psorophora* mosquitoes that are vicious biters. On the other hand, the presence of extensive mats of duckweed or salvinia (Chapter 13.4) is indicative of low mosquito production areas. Although the root system of salvinia is highly branched, this floating aquatic plant is not a preferred host for mosquito larvae.

Summary

The association between aquatic plants and certain species of mosquitoes has evolved over millions of years. The uncontrolled growth of invasive plants often provides an undisturbed habitat that mosquitoes prefer and where they can proliferate. Mosquitoes can colonize virtually any type of water body and aquatic vegetation provides a perfect environment for mosquitoes to thrive. Management of dense surface-growing exotic and native aquatic plants in permanent and temporary water systems is critical to reduce the habitats suitable for mosquito development. After all, "…Without aquatic plants, most of our freshwater mosquito problems would not exist…" (Wilson 1981).

For more information:

•Dame D and T Fasulo (eds). 2008. Public health pesticide training manual (SP318). University of Florida Cooperative Extension, Gainesville FL.
•Hoover A. 2008. First global malaria map in decades shows reduced risk. University of Florida News. http://news.ufl.edu/2008/02/26/malaria-map/
•University of Florida/IFAS Florida Medical Entomology Laboratory. Mosquito information website. http://mosquito.ifas.ufl.edu/
•O' Meara G. 2003. Mosquitoes associated with stormwater detention/retention areas. UF/IFAS Cooperative Extension Service, Gainesville FL. ENY627/ MG338. http://edis.ifas.ufl.edu/MG338.
•Wilson F. 1981. The aquatic weed-mosquito control connection. Aquatics Fall 3:6, 9-11, 14.

Photo and illustration credits:

Page 31: Mosquito life stages box (all from University of Florida IFAS Medical Entomology Laboratory)
    Upper left: *Anopheles quadrimaculatus* eggs; Roxanne Connelly
    Upper center: *Culex salinarius* larva; Michelle Cutwa-Francis
    Upper right: Mosquito pupa; James Newman
    Lower center: *Culex quinquefasciatus* adult; James Newman
Page 34: Mosquito larva attached to root of waterlettuce; T. Loyless, Florida DACS

Marc D. Bellaud
Aquatic Control Technology, Inc., Sutton MA
mbellaud@aquaticcontroltech.com

Chapter 6

Cultural and Physical Control of Aquatic Weeds

Introduction

Methods for cultural and physical control of aquatic weeds are often viewed as strategies that can be readily employed by lake users as well as lake managers. Cultural control typically focuses on education and preventing invasive species introductions from occurring in the first place. Physical control methods are usually non-chemical, non-motorized techniques that are employed to control aquatic weeds and range from hand-pulling to water-level drawdowns, or efforts to alter water or sediment characteristics where weeds are found. As awareness of aquatic nuisance species has grown in recent years, so have efforts to incorporate cultural and physical control methods as important elements of Integrated Weed Management Programs.

Prevention

Many states have prepared official lists of invasive aquatic species and some have even passed legislation to ban their transport or introduction. However, there are often limited resources or mechanisms to enforce rules and prevention efforts are often left to individual lake associations or other volunteer groups. The first step in prevention is regular monitoring to look for new or pioneer infestations. Volunteers can be trained to participate in lake monitoring or "weed-watcher" programs to accurately identify invasive species. In many cases a step-by-step reporting protocol is provided if a new "find" is discovered.

Education is a key component of prevention. Educating lake users and the general public about the threat of invasive species is necessary to prevent new infestations and to sustain effective aquatic plant management programs. Education involves creating public awareness of the problem and familiarizing people with possible solutions. Volunteer labor and public participation are paramount to successful education efforts.

Boat ramp monitoring programs are used to inspect boats and trailers for the presence of invasive species. These are largely volunteer or summer intern positions that try to staff boat ramps during peak use periods. Inspections can either be mandatory or voluntary and usually only take a matter of minutes. Several northeastern states provide annual reports about the number of "saves", which occur when an invasive species is found on a boat or trailer and is removed before the boat is launched. The interaction with boat ramp monitors also provides an opportunity to distribute educational material and conduct surveys about boating habitats and other water bodies that were recently visited.

Boat washing stations are also used at some locations as an aggressive education and prevention measure. Boats and trailers are washed prior to entering and sometimes after leaving a lake. Most aquatic plant fragments capable of surviving out of water are easily seen and can be removed by hand. Washing stations are probably better suited to removing microscopic threats such as zebra mussel veligers, didymo or spiny water flea. Primary considerations for boat washing stations are whether space and utilities for a station are available, the cost of installation, staffing and how wash water is captured and treated. Boaters are sometimes reluctant to utilize volunteer boat washing stations or those where a fee is charged and this is another hurdle that must be overcome for wash stations to be effective.

Assessment and monitoring

The accurate identification of aquatic weed infestations and their associated problems are the first steps toward developing and implementing an aquatic plant management program. Once a program is implemented, monitoring is usually warranted to evaluate the effectiveness of techniques used and to make adjustments in future years. Compliance monitoring and reporting are often a permit requirement and may focus on changes to nontarget species and water quality. The basic protocol that is recommended when initiating an aquatic plant management program is outlined in detail in Appendix D.

Physical control practices

Aeration or artificial circulation uses electric or solar powered mixers, fountains or compressed air diffuser systems to circulate and add oxygen to the water. The premise is that the addition of oxygen will reduce the amount of available phosphorus and result in less algae growth. The physical circulation or destratification (mixing) of water can also prevent noxious algal blooms from developing. Benefits of aeration have been clearly documented in all types of water bodies from small, shallow ponds to large, thermally stratified lakes that are using hypolimnetic (deep water) aeration systems. Growth of some aquatic plants appears to be limited by disturbance of the physical surface of the water and may prevent canopy formation by floating plants such as duckweed or watermeal. Recent claims that water circulators control invasive submersed species are unsubstantiated.

Benthic barriers or bottom weed barriers are used for localized control of aquatic plants through compression and by blocking sunlight. Barriers specifically manufactured for aquatic weed control are usually made from materials that are heavier than water such as PVC, fiberglass and nylon. Other fabrics used in landscaping and construction have also been tried. Barriers are usually anchored in place with a variety of fastening pins or anchoring devices. Some of the most common anchors being used are lengths of steel rebar encased in capped PVC pipes, which eliminates any sharp edges that could tear the barriers or be hazardous to swimmers. Sand bags, bricks and steel pins are also commonly used as anchors. Larger panels that are installed in water depths of greater than 4 feet usually require SCUBA divers for proper installation. Several different mechanisms have been devised to unroll the barriers in place during the installation process. Solid fabric barriers often need to be cut or vented to allow gasses to escape and to prevent billowing.

Benthic barriers are usually used to control dense, pioneer infestations of an invasive species or as a maintenance weed control strategy around boat docks and swimming areas. Large installations (greater than one acre) are often impractical due to the high cost associated with purchasing, installing and maintaining the barrier. Benthic barriers should be left in place for a minimum of 1 to 2 months to ensure that target plants are controlled, but barriers must be regularly removed and cleaned of silt; otherwise plants may begin to root on top of or through the barriers. Removal, cleaning and re-deployment is usually required every 1 to 3 years depending on the rate of silt accumulation. Some lakes with volunteer divers have attached barriers to lightweight frames that facilitate rapid deployment and retrieval. Barriers non-selectively control aquatic vegetation and may impact fish and other benthic organisms, which is another reason they are usually used for small localized areas. Many states require permits for the use of benthic barriers.

Drawdown or the lowering of the water level can be used to effectively control a number of invasive submersed species. This technique is used mostly in the northern US to expose targeted plants to freezing and drying conditions. Water is either gravity drained using a low-level gate valve or a removable flashboard system on a dam. Siphoning or pumping can also be performed in lakes with insufficient outlet structures. A principal attraction of drawdown is that it is typically an inexpensive weed control strategy for lakes with a suitable outlet structure. Annual drawdown programs can result in sediment compaction and changes in substrate composition. Drawdowns are also utilized to provide protection from ice damage to docks and other shoreline structures and to allow for shoreline clean-up and repairs by lake residents.

Plants that are usually controlled by drawdowns include many submersed species that reproduce primarily through vegetative means such as root structures and vegetative fragmentation. Some invasive submersed species most commonly targeted by drawdown include Eurasian watermilfoil (Chapter 13.2), variable watermilfoil, fanwort, egeria or Brazilian elodea (Chapter 13.8) and coontail.

Waterlily species can also be effectively controlled, provided sediments can be sufficiently dewatered to allow for the freezing and drying conditions required to control this species. Seeds and other non-vegetative propagules such as turions or winter buds are not controlled by drawdown; in fact, species that reproduce by these means may actually increase following drawdown programs. Many species of pondweed (*Potamogeton* spp.) have increased following drawdown programs and highly opportunistic species like hydrilla (Chapter 13.1) may expand rapidly following drawdown.

A general rule of thumb is to maintain drawdown conditions for 6 to 8 weeks to ensure sufficient exposure to freezing and drying conditions. Excessive snow cover or precipitation can limit the effectiveness of this technique. Drawdowns are usually timed to begin during the fall months to avoid stranding amphibians, molluscs and other benthic organisms with limited mobility. Care must also be taken to leave enough water to support fish populations and avoid impacts during key spawning periods. Drawdowns can have negative impacts on adjacent wells and wetlands as well, so it is also important to know the downstream channel configuration, capacity and flow requirements. When properly utilized, drawdowns can be a low-cost or no-cost strategy to incorporate into an integrated management program. Many states require permits for drawdown programs.

Hand pulling is one of the simplest and most widely used methods to control aquatic weed growth and can be performed by wading or from a small boat in shallow water. Snorkeling equipment or SCUBA divers are usually used in water greater than 4 to 5 feet deep and for more intensive hand pulling programs. This can be a highly selective technique, provided the target species can be easily identified. Hand pulling is usually used as a component of invasive species management programs to target new infestations with low plant density (generally less than 500 stems per acre). Hand pulling can be used to remove more dense plant growth over small areas, but benthic barriers or suction harvesting may be more effective approaches in these situations. Hand pulling is often an important follow-up strategy to a herbicide treatment program to extend the duration of plant control.

When hand pulling a plant like Eurasian watermilfoil, the roots should be carefully dislodged from the bottom substrate so that the entire plant can be collected and removed to prevent vegetative regrowth. Once the bottom substrate is disturbed, suspended sediment often greatly reduces visibility, which results in the need to make multiple passes over the same area. In larger hand pulling programs that use multiple divers, it is often advantageous to have people in boats that can collect dive bags full of weeds and can try to capture escaping plant fragments using pool skimmers.

Waterchestnut (Chapter 13.3) is a noxious invasive species that has been effectively managed in several locations by hand pulling programs. This floating-leaved plant is easily identified and is a true annual plant that usually drops its seeds in late summer. Hand pulling efforts are usually performed for several weeks during the summer months before seed drop occurs. Several successful volunteer waterchestnut hand pulling programs have been organized and implemented in the Northeast.

Hand rakes of varying sizes and configurations are being manufactured and sold for aquatic weed control. Many of these hand rakes are lightweight aluminum, with rope tethers that are designed to be thrown out into a swim area and dragged back onto shore. Some are designed to cut the weeds instead of raking them back to shore. While these may be cost-effective strategies to manage individual swim areas, there is a risk that these rakes will make the problem worse by creating weed fragments that can escape and infest other portions of the lake.

Nutrient inactivation involves the application of aluminum or iron salts or calcium compounds (lime) to remove phosphorus from the water column and to inactivate phosphorus in the sediment. Aluminum sulfate (alum) is most commonly used. Removing and inactivating phosphorus can effectively discourage algal blooms from developing, but the growth of most rooted vascular plants is usually limited by nitrogen and there are no compounds readily available that bind with nitrogen in the sediment. Injecting sediments with alum and lime has been attempted, but suppression of vascular plant growth was not significant. Nutrient inactivation remains best suited for water quality improvement and algal control (Chapter 12). In fact, reducing water column nutrients and algae may encourage even more dense infestations of nuisance rooted plants due to improved water clarity and light penetration, which may allow weeds to grow in deeper areas.

Shading through the use of EPA-registered dyes or surface covers attempts to limit light penetration and restrict the depth at which rooted plants can grow. Dyes are usually considered non-toxic solutions that give the water a blue or black color. The use of dyes is often limited to smaller golf courses or ornamental ponds because they make the water appear artificial. Dyes have little use in larger water bodies; in addition, if the pond or lake has a flowing outlet, multiple treatments may be required. Surface covers made from various fabrics or plastic materials can be used to prevent light penetration and control rooted plant growth. This approach is generally not used in recreational ponds and lakes since they would impair access to or use of the lake. Recent studies have shown that this can be an effective means of controlling plants that do not produce seeds or other vegetative reproductive propagules, but its application is usually limited to small, highly controlled areas.

Weed rollers use a roller on the lake bottom that is powered by an electric motor and travels forward and reverse in up to a 270-degree arc around a pivot point. Rollers can be up to 30 feet long and are typically installed at the end of a dock. Plants initially become wrapped around the roller and are dislodged from the sediment; the constant motion of the rollers then disrupts and compresses the bottom sediments, which prevents plants from becoming reestablished. Because the rollers travel along a pivot point, they reportedly can be used in several different substrate types. Weed rollers are only practical for managing small areas. They may disrupt fish spawning or other benthic organisms, but these impacts would likely be minimal or highly localized. Many states require permits for the use of weed rollers.

Summary

There are a number of cultural or physical methods that can be employed by lake associations or individual lakefront owners to control aquatic weeds, but the most important function of stakeholders is to develop a prevention plan. The vast majority of new weed infestations are found near boat ramps, so these areas should be surveyed on a regular basis. Residents that regularly spend time on the lake should obtain plant identification materials from state agencies or other information sources so that exotic plants can be accurately identified and targeted for treatment. Management plans should be developed for rapid response; in other words, plans should be developed proactively and stakeholders shouldn't wait for an invasive species to appear before creating a plan. Prevention and rapid response should be top priorities among lake associations because these are the most cost-effective and ecologically sound means of protecting aquatic resources from invasive species.

For more information:
- Center for Aquatic and Invasive Plants. University of Florida IFAS website. http://plants.ifas.ufl.edu/
- Krischik VA. Managing aquatic plants in Minnesota lakes. University of Minnesota Extension website. http://www.extension.umn.edu/distribution/horticulture/DG6955.html
- Lembi CA. Aquatic plant management. Purdue University. http://www.ces.purdue.edu/extmedia/ws/ws_21.pdf
- Maine Department of Environmental Protection Invasive Aquatic Plants website. http://www.state.me.us/dep/blwq/topic/invasives/
- New Hampshire Department of Environmental Services, Volunteer Weed Watcher Program website. http://des.nh.gov/organization/divisions/water/wmb/exoticspecies/weed_watcher.htm
- Vermont Department of Environmental Conservation, Water Quality Division, Aquatic Invasive Species website. http://www.vtwaterquality.org/lakes/htm/ans/lp_ans-index.htm
- Wagner KJ. 2004. The practical guide to lake management in Massachusetts. http://www.mass.gov/dcr/waterSupply/lakepond/downloads/practical_guide.pdf
- Washington Department of Ecology, Aquatic Plant Management website. http://www.ecy.wa.gov/programs/wq/plants/management/index.html

Photo and illustration credits
Page 36 upper: Boat wash station; William Haller, University of Florida Center for Aquatic and Invasive Plants
Page 36 lower: Benthic barrier; William Haller, University of Florida Center for Aquatic and Invasive Plants
Page 37: Drawdown; University of Florida Center for Aquatic and Invasive Plants (photographer unknown)

William T. Haller
University of Florida, Gainesville FL
whaller@ufl.edu

Chapter 7

Mechanical Control of Aquatic Weeds

Introduction

The term "mechanical control" as used in this chapter refers to control methods that utilize large power-driven equipment. The simplest method of mechanical control might be the dragging of an old bedspring or other heavy object behind a boat to rip up and remove submersed weeds from a beach used for swimming. Mechanical control has been practiced in the US for over a century and almost every engineer has a conceptual idea on how to build the "perfect aquatic weed harvester." One major obstacle to designing a universal mechanical harvester is the diversity of plants and environments where the equipment will be employed. This has led to the development – and ultimate abandonment – of a plethora of various types of equipment throughout the years. Primary factors to be considered when selecting a mechanical control method are the types of weeds to be controlled and habitats they occupy.

Wetland or emergent weeds

Wetland habitats are typical marsh ecosystems with periodically inundated soils, a high water table and/or water depths of up to two feet. Emergent plants such as phragmites (Chapter 13.9), purple loosestrife (Chapter 13.6), cattails and other wetland plants are common in these areas. Mechanical control is employed on a very limited basis in these "protected" habitats because access is often difficult and the destruction and alteration of protected wetlands in the US is highly regulated.

While there is very limited mechanical weed control conducted in wetlands, the mechanical control method most commonly employed by land managers is mowing. For example, dense stands of phragmites may be mowed during dry seasons or under drought conditions to provide temporary control. Also, chain saws and hand-pulling have been used in wetlands of southern Florida for control of melaleuca trees and seedlings, respectively. Ducks Unlimited and other resource agencies have used dredges and choppers of various types to reclaim or restore wetlands, but the primary purpose of these activities is not solely weed control. Overall, mechanical weed control is rarely used for invasive species management in wetlands and shallow-water areas due to the likelihood of creating significant environmental damage.

Floating weeds

Most mechanical weed control occurs in water greater than 2 feet deep and the type of plant to be controlled (floating or submersed) must be taken into consideration when selecting a mechanical control method. Floating plants should be evaluated separately from submersed plants because floating plants produce 10 to 20 times more biomass than submersed plants – biomass that has to be chopped, picked up or otherwise moved away from the harvesting site. For example, the standing crop or biomass of an acre of undisturbed waterhyacinth (Chapter 13.5) can weigh 200 to 300 tons per acre, whereas an acre of hydrilla (Chapter 13.1) or Eurasian watermilfoil (Chapter 13.2) can weigh only 10 tons or less per acre. Most mechanical harvesters are able to pick up and transport less than 5 tons of biomass per load, so there is a huge difference in the time, effort and expense required to mechanically harvest floating plants compared to submersed aquatic weeds.

Two additional problems associated with floating plants are their ability to move by wind or water currents and their location in lakes and rivers. For example, there may be only one access point where plants can be loaded onto trucks for disposal. Plants may initially be located close to the work site, but on another day – after a change in wind direction – plants may be on the other side of the lake and will need to be transported a long distance before they can be off-loaded. Also, floating plants are often blown into shallow waters along shorelines, which may be lined with cypress or willow trees. Most harvesters cannot work in water less than 2 feet deep and cannot navigate in and among trees, rocks or stump-fields in flooded reservoirs.

Submersed weeds

Mechanical harvesting of submersed weeds, primarily curlyleaf pondweed (Chapter 13.7) and Eurasian watermilfoil, has been utilized in the Northeast and Midwest. The shallow shores of even very deep lakes in these regions often support the growth of these submersed weeds and multiple harvests provide control during the recreational season. Governmental entities (including state, county and local governments) have subsidized weed removal from public lakes in some locations to maintain high use areas and to promote tourism and general utilization of the water resource. In other areas, lake associations and groups of homeowners often hire aquatic management companies for weed removal services. Although mechanical harvesting is often used in northern lakes to control submersed weeds, this method has less utility in southern states due to longer growing seasons and much larger-scale coverage of weeds in the shallow lakes and reservoirs more commonly encountered in the Southeast.

Examples of mechanical equipment

Cutter boats have been used in the US in one form or another for decades. For example, a small barge with a steam engine powered an underwater sickle bar mower in the Upper Chesapeake Bay/Potomac River area at the turn of the century. Submersed plants cut by the barge floated from the harvested area via river and tidal currents. Also, the US Army Corps of Engineers built sawboats in the early 1900s for use in navigable waters of Louisiana and Florida. These boats had gangs of circular saws mounted about an inch apart on a spinning shaft that was mounted at the bow of the boat and only penetrated the top inch or two of the water. These sawboats chopped up waterhyacinth, alligatorweed and grasses which formed intertwined mats of vegetation. The chopped vegetation was allowed to flow downstream or to salt water. Cutter boats have been used more recently to clear navigation channels, but this equipment is not usually used in lakes and non-flowing systems because most cut weeds float and survive for long periods of time. Fragments such as these can establish in other parts of the water body or wash up on swimming beaches. Cutter boats create large amounts of fragments and vegetative cuttings, so the ability of the target weed to spread and grow from fragments should be evaluated before cutter boats are employed as a primary mechanical control method.

Shredding boats are used to control emergent and floating plants. The most common type of shredder is the "cookie cutter," which consists of two spinning blades (3 to 4 feet wide) that are mounted behind a steel hood on the front of a small but powerful barge. The boat is propelled by hydraulically raising and lowering the blades and changing the direction of the blades (see www.texasharvesting.com). Recently, bow mounted high-speed flail mower blades have been tested for chopping and shredding floating and emergent plants. As with other mechanical control equipment, shredder boats are very specialized pieces of equipment, are non-selective and create many plant fragments. However, they work well when used in the areas for which they are designed and are frequently used in wetland restoration projects, where removal of cut vegetation is too expensive or not feasible.

Rotovators are highly specialized large aquatic rototillers. The rotovator head is lowered into the lake or river bottom and "tills" the sediments, which chops up and cuts loose submersed plants. A floating boom is usually placed around the work area while the rotovator spins on the lake bottom; uprooted plants float to the surface and are removed from along the barrier by hand or mechanical means. Rotovators have been used mostly in the Pacific Northwest, where the submersed weed Eurasian watermilfoil grows in rocky bottom areas and roots in the shallow soil between and among small rocks. The rotovator head moves the rocks around and uproots the weeds from the shallow soils and rock crevasses.

Dredges are not usually used for aquatic weed control due to high costs associated with their operation, but weed control can be a benefit of dredging that is done for other reasons. Shallow ponds and lakes that have filled with silt and organic matter over time may only be 3 to 4 feet deep and provide an ideal environment for excessive growth of submersed weeds and native plants such as cattail and waterlily. If the water depth of the pond is increased to 6 to 10 feet by dredging, it is unlikely that emergent plants such as cattail will continue to grow. However, submersed weeds will almost certainly still infest the pond if water depth and clarity requirements for growth of the weeds are met.

Harvest and removal harvesters are the most widely used types of equipment employed for mechanical control in the US. The first machines were developed in the 1950s by a Wisconsin company to harvest Eurasian watermilfoil and curlyleaf pondweed from the edges of the hundreds of lakes in the Upper Midwest. These lakes are generally deep in the middle and aquatic weeds naturally grow in the shallow littoral areas, which receive intensive use for swimming and docking. Harvest and removal harvesters are powered by side-mounted paddle wheels which operate independently in forward or reverse. As a result, these harvesters are highly maneuverable around docks and boat houses. Also, the machines can operate in as little as 12 to 18 inches of water. These harvesters cut plants off at depths of 5 feet and in swaths 8 feet wide with a hydraulically operated cutter head and convey the cut plants into a storage bay on the harvester. When the harvester is full, it offloads harvested plants onto a transport barge by conveyer belts and the transporter takes the vegetation to shore, where it is dropped onto a conveyor to elevate the load to a truck for disposal. If you have read this carefully, you have counted four pieces of equipment: a harvester, a transporter, a shore conveyer and a truck. All this equipment may not be necessary, as mechanical harvesting is obviously tailored to a particular situation and is very site-specific. Also, some harvester trailers have been modified to allow them to transport cut weeds to the disposal site. This system or a setup with similar equipment has been used for 50 years in lakes from New England to California, but is mostly employed in northern lakes where one or two harvests during spring and summer can provide weed-free conditions for the seasonal summer use of these lakes.

Advantages and disadvantages

There are many advantages to mechanical harvesting. These include:

- Water can be used immediately following treatment. Some aquatic herbicides have restrictions on use of treated water for drinking, swimming and irrigation. Also, plants are removed during mechanical harvesting and do not decompose slowly in the water column as they do after herbicide application. In addition, oxygen content of the water is generally not affected by mechanical harvesting, although turbidity and water quality may be affected in the short term.

- Nutrient removal is usually insignificant because only small areas of lakes (1 to 2%) are typically harvested; however, some nutrients are removed with the harvested vegetation. It has been estimated that aquatic plants contain less than 30% of the annual nutrient loading that occurs in lakes.

- The habitat remains intact because most harvesters do not remove submersed plants all the way to the lake bottom. Like mowing a lawn, clipped plants remain rooted in the sediment and regrowth begins soon after the harvesting operation.

- Mechanical harvesting is site-specific because plants are removed only where the harvester operates. If a neighbor wants vegetation to remain along his or her lakefront, there is no movement of herbicides out of the intended treatment area to damage the neighbor's site.

• Herbicide concerns remain widespread despite extensive research and much-improved application, use and registration requirements that are enforced by regulatory agencies (Appendix A). Mechanical harvesting, despite some environmental concerns (as outlined below), is perceived to be environmentally neutral by the public.

• Utilization of harvested biomass is thought by many to be a means of offsetting the relatively high costs and energy requirements associated with mechanical harvesting. Unfortunately, no cost-effective uses of harvested vegetation have been developed, despite much research examining the utility of harvested plant material as a biofuel, cattle feed, soil amendment, mulch or even as a papermaking substrate. As much as 95% of the biomass of aquatic plants is water, so 5 tons of Eurasian watermilfoil yields only 500 pounds of dry matter. In addition, cut plants in northern lakes are only available for 3 to 4 months of the year.

The easiest way to highlight the disadvantages of mechanical harvesting is to point out that major producers of farm equipment (for example, John Deere or New Holland) do not mass-produce equipment designed for the mechanical harvesting of aquatic weeds. Farmers are famous for efficiently cutting, harvesting and moving hay, corn and grain crops; they constitute a large market and specialized equipment is available to them. On the other hand, the demand for aquatic weed harvesters is very small, so the equipment associated with these operations is often custom-made and expensive. Other disadvantages include:

• The area that can be harvested in a day depends on the size of the harvester, transport time, distance to the disposal site and density of the weeds being harvested. These factors can result in a wide range of costs. The cost of harvesting is site-specific, but mechanical harvesting is generally more expensive than other weed control methods due to the variables noted above and the generally high capital outlay required to purchase equipment that may only be used for 3 or 4 months per year.

• Mechanical harvesters are not selective and remove native vegetation along with target weeds. However, this is probably not a significant disadvantage since native plants and weeds will likely return by the next growing season, if not sooner.

• By-catch, or the harvesting of nontarget organisms such as fish, crayfish, snails and frogs along with weeds, may be more of a concern, but the degree or extent of harvesting should be considered. Research on fish catch during mechanical harvesting of submersed vegetation has shown that 15 to 30% of some species of fish can be removed with cut vegetation during a single harvest. If the total area of a lake that is harvested is 1, 5 or 10% of the lake's area, this will likely be of little consequence. However, if the management plan for a 10-acre pond calls for complete harvests 3 times per year, then the issue of by-catch of fish deserves more consideration.

• Regrowth of cut vegetation can occur quickly. For example, if hydrilla can grow 1" per day as reported, a harvest that cuts 5 feet deep could result in plants reaching the water surface again only two months after harvesting. Speed of regrowth depends on the target weed, time of year harvested, water clarity, water temperature and other factors.

• Floating plant fragments produced during mechanical harvesting can be a concern because most aquatic weeds can regrow vegetatively from even small pieces of vegetation. If an initial infestation of aquatic weeds is located at a boat ramp, care should be taken to minimize the spread of fragments to uninfested areas of the lake by maintaining a containment barrier around the area where mechanical harvesting will take place. On the other hand, if a lake is already heavily infested with a weed, it is unlikely that additional fragments will spread the weeds further. However, homeowners downwind of the harvesting site may not appreciate having to regularly rake weeds and floating fragments off their beaches.

• Disposal of harvested vegetation can be an expensive and difficult problem after mechanical harvesting. Research during a project in the 1970s on Orange Lake in Florida compared the costs of in-lake disposal to the transport, off-loading and disposal of cut material at an upland site. As water levels on Orange Lake decreased during a drought period, the mechanical harvester was allowed to off-load cut vegetation along the shoreline among emergent vegetation instead of transporting harvested plants to the shore for disposal. The cost of in-lake disposal reduced the per-acre cost by about half when compared to transporting the vegetation to shore, loading it into a truck and disposing of the plant material in an old farm field.

• Some lakes or rivers may not be suitable for mechanical harvesting. If there is only one public boat ramp on a lake and it is not close to the area to be harvested, the costs of moving the cut vegetation from the harvester to shore will add significantly to the cost of the operation. Harvesters are not high-speed machines and move at 3 to 4 mph, so if a river flows at 2 mph and the harvester has to travel upstream to the off-loading site, well, do the math! Off-loading sites usually must have paved or concrete surfaces because the weeds are wet and an unpaved off-loading site can quickly become a quagmire.

Summary

This discussion is not intended to include all the machines that are available for mechanical control of aquatic weeds and it is likely that new ideas and equipment will be developed as time passes. It is important to remember that each site and each weed has characteristics that may require a particular type of mechanical harvester and may preclude the use of other mechanical methods of control. There is a vast repository of information available on the internet and the best source of information is the conservation or regulatory agency in your state. In fact, most states require that permits for mechanical harvesting be obtained before work can begin. For a further discussion of mechanical control, photos of equipment and a list of equipment manufacturers, please visit http://plants.ifas.ufl.edu/guide/mechcons.html.

For more information:
- http://www.ecy.wa.gov/Programs/wq/plants/management/aqua026.html
- http://www.co.thurston.wa.us/stormwater/Lakes/Long%20Lake/Long_Harvesting.htm

Photo and illustration credits:
Page 41: Mower; William Haller, University of Florida Center for Aquatic and Invasive Plants
Page 42: Cutter boat; William Haller, University of Florida Center for Aquatic and Invasive Plants
Page 43: Harvester; William Haller, University of Florida Center for Aquatic and Invasive Plants
Page 44: Conveyor; Jeff Schardt, Florida Fish and Wildlife Conservation Commission

Chapter 8

James P. Cuda
University of Florida, Gainesville FL
jcuda@ufl.edu

Introduction to Biological Control of Aquatic Weeds

Introduction

There are many herbivores or plant-eating animals in the aquatic environment, including moose, muskrat, turtles, fish, crayfish, snails and waterfowl. These animals are general herbivores and may prefer to eat certain types of plants, but do not rely on a single plant species as a primary food source. Although these animals do consume aquatic plants and therefore reduce the growth of some species, they generally do not have a significant impact on overall plant growth because they feed on many different plants and are not considered biological control agents. Biological control (also called biocontrol) is broadly defined as the planned use of one organism (for example, an insect) to control or suppress the growth of another organism such as a weedy plant species. Biocontrol of weeds is primarily the search for, and introduction of, species-specific organisms that selectively attack a single target species such as an exotic weed. These organisms may be insects, animals or pathogens that cause plant diseases, but most biocontrol agents are insects. Biocontrol has been studied and used for more than a century and has developed into a complicated and technical science based on a number of principles that will be discussed in this chapter. Two different approaches are currently used in the biocontrol of aquatic weeds: classical (importation) and non-classical (augmentation, conservation).

Classical biocontrol is by far the most common biological control method and typically involves the introduction of natural enemies from their native range to control a nonnative invasive plant. The excessive growth of a weed in its new habitat is due in part to the absence of natural enemies that normally limit or slow the growth, reproduction and spread of the weed in its native range. Classical biocontrol seeks to reunite an invasive plant with one or more of its coevolved natural enemies to provide selective control of the weed. Thus, classical biocontrol can be defined as the planned introduction and release of nonnative target-specific organisms (usually arthropods, nematodes or plant pathogens) from the weed's native range to reduce the vigor, reproductive capacity or density of the target weed in its adventive (new or introduced) range.

Classical biocontrol offers several advantages over other weed control methods. It is relatively inexpensive to develop and use compared to other methods of weed control. Classical biocontrol provides selective, long-term control of the target weed and because biocontrol agents reproduce, they will usually spread on their own throughout the infested area. Some of the strengths of classical biocontrol also contribute to its shortcomings. For example, it may not be possible to find a biocontrol agent that effectively controls a single weed and selectively attacks only that particular weed. When potential biocontrol agents are identified, their establishment and suppression of the target weed in the introduced area are not guaranteed. Even if biocontrol agents do successfully establish in their introduced areas, control is not immediate and agents may require many years to have a major impact on target weeds. Finally, once a biocontrol agent is established, it cannot be recalled if desirable nontarget species are affected by the agent.

Non-classical biocontrol involves the mass rearing and periodic release of resident or naturalized nonnative aquatic weed biocontrol agents to increase their effectiveness. Savvy home gardeners employ this approach when they purchase ladybird beetles to control aphids (insects that are serious pests of fruits, vegetables and ornamentals) in their home gardens. Augmentative or repeated releases of native or naturalized insects have occasionally been used for suppression of alligatorweed, waterhyacinth (Chapter 13.5), hydrilla (Chapter 13.1) and Eurasian watermilfoil (Chapter 13.2).

The "new association" approach is a variation of classical biocontrol. New association biocontrol differs from classical biocontrol in that the natural enemies or biocontrol agents have not played a major role in the evolutionary history of the host plant and are therefore considered new associates. Because organisms used in the new association approach are not entirely host-specific, this approach is appropriate only in cases where the target weed has few or no closely related native relatives in the area of introduction.

A good example of the new association approach is the milfoil weevil (*Eurychiopsis lecontei*), which is native to North America and attacks native species of milfoil (*Myriophyllum* spp.) in the US and Canada. Recent studies have shown that milfoil weevils reared on the introduced weed Eurasian watermilfoil (*Myriophyllum spicatum*) not only develop faster and survive better on the exotic invasive milfoil, but also preferentially attack the nonnative weed species over the native northern watermilfoil (*M. sibiricum*), its natural host plant. This phenomenon was unexpected, unplanned and unusual. Many aquatic resource managers are currently evaluating this natural occurrence to determine how best to include this weevil in weed control programs.

Procedures in a Classical Weed Biocontrol Project

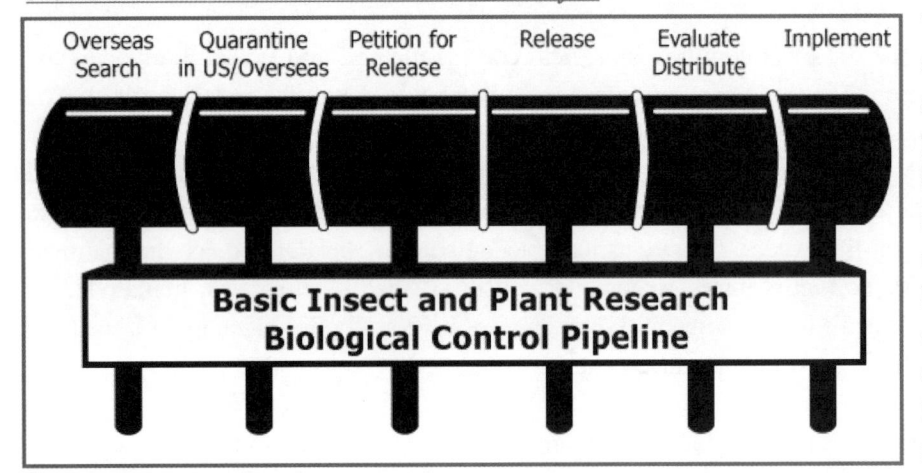

Weed biocontrol scientists (most of whom are entomologists or pathologists) develop and refine procedures for locating, screening, releasing and evaluating biocontrol agents. All countries currently conducting weed biocontrol projects follow this protocol in one form or another to ensure that candidate organisms are safe to introduce. The normal process in a classical biocontrol program is often referred to as the "pipeline." The pipeline consists of the following series of well-defined steps:

Step 1: Target selection. Ideal targets for biocontrol are invasive nonnative aquatic plants with no closely related native plants in their introduced ranges. Scientists read the literature associated with the target weed to learn where the weed came from (geographic origin), what desirable plants are closely related to the weed and to identify potential natural enemies.

Step 2: Overseas and domestic surveys. Scientists visit the native range of the target weed to search for natural enemies that may affect and slow the growth of the weed. They evaluate how the target weed is damaged by organisms in its native range to determine if these organisms may be useful as biocontrol agents for the target weed in its introduced range. Another predictor of success is past performance; if a biocontrol agent has been successful in controlling a weed in some countries, there is a high probability that it will be successful in other countries as well. Scientists also conduct surveys in the weed's introduced range (domestic surveys) to avoid introducing biocontrol agents that are already established but ineffective.

Step 3: Importation and quarantine studies. If an organism attacks only the exotic weed, and not desirable species, scientists request permission from the US Department of Agriculture to import the organism to the US for host range testing. Once permission for importation is granted, the potential biocontrol agent is brought to the US and placed in an approved quarantine laboratory where it cannot escape and is carefully studied to ensure it will not harm desirable species such as crops and native plants.

Step 4: Approval for release. The results of quarantine studies are forwarded to the appropriate federal and state agencies, who determine whether the organism is safe to release. These independent agencies may request that additional testing be done to evaluate the effect of the organism on additional native plants, especially threatened or endangered species, as well as related plants not included in the original quarantine studies.

Step 5: Release and establishment. Once the biocontrol agent is shown to pose minimal risks to desirable native, ornamental and crop plants, permits are issued and large numbers of the biocontrol agent are reared. This ensures that population densities will be high enough to allow breeding colonies of the agent to establish in the field. Scientists then release the biocontrol agent in multiple locations to increase the likelihood of successful establishment.

Step 6: Evaluation. Scientists monitor all introduced biocontrol agents after field release to confirm establishment and dispersal of the agent. Multiple releases of the organism may be necessary initially to maintain populations that are adequate for control of the weed species. Studies are also conducted to determine the effect of the biocontrol agent on the target weed as well as on additional nontarget plants.

Step 7: Technology transfer. Resource managers are trained in the identification and use of the biocontrol agent. Scientists also collaborate with those using the biocontrol agent to determine the best methods to integrate biocontrol with other weed control methods.

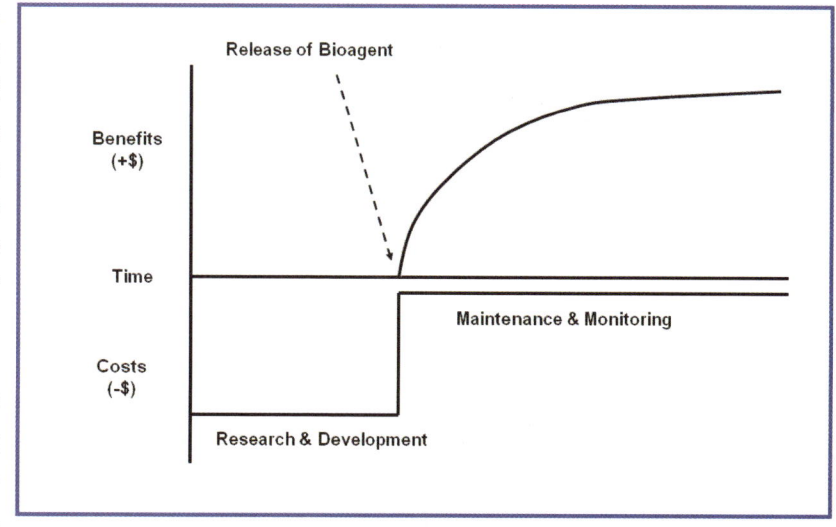

Successful biocontrol programs are expensive at the beginning and can take a long time to develop, but biocontrol can reduce the need for other weed control methods such as herbicides and mechanical harvesting. Because classical biocontrol can provide selective, long-term control of a target weed and biocontrol agents naturally spread by reproducing, the use of biocontrol results in the reduction or elimination of costs for other aquatic weed control methods.

Safety – what has to be done to introduce a biocontrol agent?
Host specificity is fundamental to biological weed control because it ensures that an introduced agent will not damage desirable plants. Host-specific, coevolved natural enemies are considered good candidates for use as biocontrol agents because they are unable to reproduce on plants other than their weedy hosts. In addition, these types of organisms have proven to be the safest to introduce because they are least likely to damage nontarget species. Because host-specific natural enemies reproduce only when they have access to their host plants, their populations are limited by availability and abundance of the target weed.

Potential biocontrol agents are first tested for effectiveness and host specificity in their native range, then promising candidates are brought to quarantine laboratories in the US for final host range testing to determine whether the organism can live and reproduce on native plants. Before scientists can release an agent into the US for classical biocontrol of an invasive aquatic plant, the potential agent must undergo rigorous testing in quarantine to ensure it will only survive on the weed species and will not harm nontarget species. The potential biocontrol agent is offered a series of carefully chosen plants in two different types of tests to determine if the agent is safe to release. In no-choice tests, the agent is given access only to a nontarget plant to determine if it will attack the nontarget plant if the agent's host plant (the target weed) is unavailable. In multiple-choice tests, the agent is offered the target weed and at least one nontarget plant to determine whether the agent damages only the target weed. <u>Nonnative biocontrol agents can only be released if these tests show that the agent requires the host plant to survive and reproduce and that it will not attack desirable nontarget plants.</u>

Selecting organisms as candidates for classical biocontrol is a complicated and lengthy process because scientists must identify natural enemies that have developed a high degree of specificity with their weedy host plants. According to established guidelines, no potential biocontrol agent can be introduced into a new environment before its host range is determined. Multiple screening tests are usually required to identify the host range of the agent and scientists must conduct a number of host range tests in the field and laboratory (egg laying, larval development and feeding by adults) to determine whether a biocontrol agent requires the presence of the weedy host plant to survive. Candidate organisms that are able to live and reproduce without access to their weedy host fail the host specificity requirement; they are then dropped from further consideration and quarantined populations are destroyed.

The review process – why does it take so long to release a biological control insect?
The US Department of Agriculture's Animal and Plant Health Inspection Service, Plant Protection Quarantine permitting unit (hereafter referred to as APHIS) is responsible for approving the release of any biocontrol agent in the US. The Plant Protection Act of 2000 gives APHIS the authority to regulate "any enemy, antagonist or competitor used to control a plant pest or noxious weed." Scientists must apply for a permit from APHIS before they can import a potential biocontrol agent into the US for host specificity testing and approved biocontrol agents must be sent directly to a high-security quarantine facility upon entry into the US. There are a number of secure quarantine facilities located throughout the US that are specifically designed and constructed for biocontrol research on aquatic and terrestrial weeds.

After host specificity testing is completed, a permit must be obtained from APHIS before the biocontrol agent is released in the field. A multi-agency Technical Advisory Group for Biological Control Agents of Weeds (TAG) reviews information submitted by the requesting scientist to APHIS. TAG members review test plant lists for weed biocontrol projects, advise weed biocontrol scientists, review petitions for field release of weed biocontrol agents and provide APHIS with recommendations on the proposed release.

In addition to submitting a release petition to TAG and APHIS, scientists contact the Department of the Interior to ensure that threatened and endangered species are included in their test plant list. Release of nonnative weed biocontrol agents also requires compliance with the Endangered Species Act (ESA) and the National Environmental Policy Act (NEPA). Scientists must complete an Environmental Assessment (EA) document that outlines the potential impact of the biocontrol agent on the environment in order to comply with the NEPA. The EA provides the public with possible positive and negative environmental impacts that might occur if the new biocontrol agent is released in the US. Scientists must also submit to the US Fish and Wildlife Service a Biological Assessment (BA) document in order to comply with the ESA. The review process is designed in this manner to ensure that there is little chance the introduced biocontrol agents will become pests themselves. Once a weed biocontrol agent is released, several years may be required for the organism to establish and impact the target weed. Scientists continually monitor dispersal of the agent, collect data on its effectiveness to the target weed and also monitor the agent's effect (if any) on nontarget plants during this time.

What is considered a success?

Successful biocontrol of an aquatic weed is a function of the biocontrol agent's capacity to reproduce on individual plants and to build populations large enough to damage the weed's population. However, high population densities of a biocontrol agent do not necessarily guarantee success and effective biocontrol may only occur when the weed is stressed concurrently by local climatic conditions, competing plants or other natural enemies.

In general, insect biocontrol of aquatic weeds in the US has been successful since it was first used to control alligatorweed in 1964. Insects have provided varying levels of control (from complete control to suppression of growth) of the aquatic form of alligatorweed and of waterhyacinth in most areas where insect biocontrol has been attempted. The high success rate achieved by these projects may be correlated with the growth form of the weeds, their susceptibility to disease-causing pathogens, the fluid nature of the aquatic environment, the organisms used as biocontrol agents, or a combination of these factors. For instance, waterhyacinth and the aquatic form of alligatorweed produce floating mats, a growth habit that makes them susceptible to wave action and currents that are unique to aquatic environments. Also, reproduction of these weeds is due primarily to rapid vegetative growth, which results in clonal populations with little or no genetic diversity. Since many plant defenses against diseases and insects (including biocontrol agents) are determined by the genetic composition of a plant, the entire population of a clonally reproducing species would likely react to a biocontrol agent in the same manner; that is, if one plant is damaged by the biocontrol agent, the entire population is likely to be damaged by the agent as well. Waterhyacinth and the aquatic form of alligatorweed also are highly susceptible to secondary infection, so plants that have been injured by insects or disease rot and disintegrate very rapidly. Finally, beetles – especially weevils – have been responsible for most successful biocontrol programs. Adults of these insects tend to remain above the water, which may reduce fish predation, whereas larvae often feed inside the plant. These habits allow them to maintain high density populations in the environment. A number of successful weed biocontrol programs have utilized members of the insect group Coleoptera; in fact, the majority (greater than 75%) of insects released thus far for biocontrol of aquatic plants are weevils and beetles.

Defining success in biocontrol of weeds is usually subjective and highly variable. A project may be considered successful in an ecological sense when a biocontrol agent successfully establishes in an area and reduces the target weed's population. However, the severity of damage inflicted by the biocontrol agent may not result in the level of control desired by lake managers, boaters and homeowners. Recently, a clear distinction has been made between "biological success" and "impact success." Biocontrol agents can be biologically successful (they establish and sustain high population densities on the target weed), but may not realize impact success (they do not provide the desired level of control or impact on the weed).

The use of terms that define success (such as complete, substantial or negligible) in a biocontrol program may not take into account variations in time and space. For example, in the southeastern United States where the alligatorweed flea beetle has been introduced, biocontrol success can range from complete to negligible depending on the season, geographic area and habitat (Chapter 9). However, these terms can be useful from an operational perspective since they describe the current success level of biocontrol efforts and help managers to determine which other control measures (e.g., harvesters, aquatic herbicides) must be used to achieve the desired level of weed control. The advantage of this system is that it describes success in practical terms that are more readily understood by aquatic plant managers and the public. For example, biocontrol is defined as *complete* when no other control method is required, *substantial* when other methods such as herbicides are still required but at reduced levels and *negligible* when other control methods must be used at pre-biocontrol levels to manage the weed problem.

Summary

Biocontrol historically has been a major component of integrated pest management programs for terrestrial insect and weed control and can be an effective tool in the aquatic weed manager's arsenal as well. Classical biocontrol, which relies on importation of natural enemies from a weed's native home, may be useful to control an exotic invasive species that thrives when introduced to an area that lacks the natural enemies responsible for keeping the weed in check in its native range. The use of host-specific biocontrol agents allows management of populations of weedy species while leaving nontarget native plants unharmed. Successful biocontrol programs are often expensive and time-consuming to develop, but if successful can provide selective, long-term control of a target weed. Although a number of types of organisms – including disease-causing plant pathogens, insects and grass carp – have been studied for potential use as biocontrol agents, the greatest successes in aquatic systems have been realized with insects (Chapter 9) and grass carp (Chapter 10).

For more information:
- Biological control for the public
http://everest.ento.vt.edu/~kok/Text_frame1.htm
- Biological control of weeds – it's a *natural*!
http://www.wssa.net/Weeds/Tools/Biological/BCBrochure.pdf
- Biological control of weeds: why does quarantine testing take so long?
http://ipm.ifas.ufl.edu/applying/methods/biocontrol/quarantinetest.shtml
- Harley KLS and IW Forno. 1992. Biological control of weeds: a handbook for practitioners and students. Inkata Press, Melbourne, Australia.

- How scientists obtain approval to release organisms for classical biological control of invasive weeds
http://edis.ifas.ufl.edu/IN607
- Mentz KM. 1987. The role of economics in the selection of target pests for a biological control program in the South-west Pacific, pp. 69-85. In P Ferrer and DH Stechman (eds.), Biological control in the South-west Pacific. Report on an international workshop, Vaini, Tonga, 1985. Government Printing Office, Tonga.
- Plant management in Florida waters – biological control
http://plants.ifas.ufl.edu/guide/biocons.html
- Smith RF and R van den Bosch. 1967. Integrated control, pp. 295-340. In WW Kilgore and RL Doutt (eds.). Pest control – biological, physical and selected chemical methods. Academic Press, New York.
- US Army Corps of Engineers Aquatic Plant Information System
http://el.erdc.usace.army.mil/aqua/APIS/apishelp.htm

Photo and illustration credits:

Page 48: Biocontrol pipeline; Joshua Huey, University of Florida Center for Aquatic and Invasive Plants
Page 49: Biocontrol graph; Harley and Forno, 1992

Chapter 8

James P. Cuda
University of Florida, Gainesville FL
jcuda@ufl.edu

Chapter 9

Insects for Biocontrol of Aquatic Weeds

Introduction

Biocontrol of aquatic weeds with insects has resulted in the successful establishment of many potential biocontrol insects since it was first attempted in the US against alligatorweed in 1964. Aquatic weeds have historically been a more serious problem in the southern US due to the moderate climate and shallow lakes in these regions where weeds often cover large areas. Consequently, the greatest body of research on biocontrol has focused on weeds of the southern US. The following section describes in detail the relationship of particular biocontrol insects introduced into the US and their target weeds.

Alligatorweed (*Alternanthera philoxeroides*)
(see http://el.erdc.usace.army.mil/aqua/APIS/apishelp.htm for more information)

Enemy	Type	Origin (Date)	Success	Comments
Agasicles hygrophila	Beetle	Argentina (1964)	Complete (south); Negligible (north)	Found throughout the southern 2/3 of the range of alligatorweed in the US where it provides almost complete control
Amynothrips andersoni	Thrips	Argentina (1967)	Negligible	Attacks terrestrial plants more than the other species
Arcola (=*Vogtia*) *malloi*	Moth	Argentina (1971)	Negligible	Most important control agent in the upper Mississippi valley

If this manual had been written in the 1960s and 1970s, alligatorweed (introduced in the late 1800s) would have been included as one of the worst weeds in the US in Chapter 13. The alligatorweed flea beetle (*Agasicles hygrophila*) was introduced in 1964 and has provided excellent control of the floating form of alligatorweed from southern Florida along the Gulf Coast to southern Texas. Unfortunately the alligatorweed flea beetle is not as cold-tolerant as alligatorweed and insect populations die out during severe winters in the central and northern parts of the Gulf states.

Alligatorweed remains a problem in areas such as central and northern Texas, Mississippi, Alabama, Georgia and the Carolinas. The alligatorweed flea beetle is self-sustaining in its southern range but not in the north. The US Army Corps of Engineers periodically collects and re-releases the beetle in northern areas during spring to reestablish northern populations. This is an example of combining augmentation with classical biocontrol. The alligatorweed flea beetle has eliminated the need for other forms of control in natural areas when it is well-established. The *Amynothrips* and *Arcola* insects also are established on alligatorweed; the *Amynothrips* attacks the terrestrial alligatorweed plants more than do the other species. However, control of alligatorweed is largely attributed to the alligatorweed flea beetle.

Alligatorweed provides a good example of how a biocontrol agent controls its weedy host plant without completely eradicating the population of the weed. Alligatorweed grows quickly in spring and populations of the alligatorweed flea beetle increase as well, but lag behind development of the host plant. By the time alligatorweed has grown enough to become problematic, the population of the

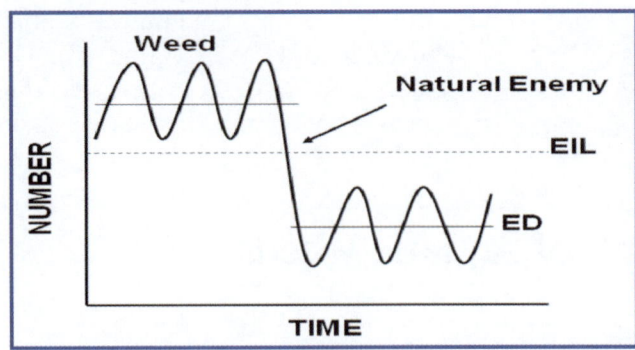

EIL = economic injury level; ED = equilibrium density

alligatorweed flea beetle reaches a density sufficient to destroy most of the alligatorweed. The number of alligatorweed flea beetles then decreases, alligatorweed growth resumes and the cycle begins anew. This is a nearly perfect example of a highly successful insect biocontrol program that adequately controls an invasive aquatic plant. Furthermore, 40 to 50 years after their introduction, none of the three insects released to control alligatorweed have been found feeding on, reproducing on or otherwise affecting nontarget native species.

Waterhyacinth (Chapter 13.5)
(see http://el.erdc.usace.army.mil/aqua/APIS/apishelp.htm for more information)

Enemy	Type	Origin (Date)	Success	Comments
Neochetina bruchi	Weevil	Argentina (1974)	Substantial	Widely distributed throughout the range of waterhyacinth in the US
Neochetina eichhorniae	Weevil	Argentina (1972)	Substantial	
Niphograpta albiguttalis	Moth	Argentina (1977)	Negligible	Prefers plants with short bulbous petioles
Orthogalumna terebrantis	Mite	USA (native)	Negligible	Produces characteristic dark stripes in the leaves; also attacks pickerelweed

Two *Neochetina* weevils and the *Niphograpta* stem-boring caterpillar have been released as biocontrol agents of waterhyacinth. The life cycle of the *Neochetina* weevils requires about 2 to 3 months to complete and is dependent on temperature. These weevils act on waterhyacinth by causing feeding damage that reduces the plant's ability to regenerate. Adult weevils produce characteristic rectangular feeding scars on the leaves, whereas larvae tunnel inside the leaf petioles to the crown or meristem where they damage new growth. Feeding damage also allows plant pathogens to invade the feeding scars and larval tunnels, which further weakens the plant. The life cycle of the *Niphograpta* caterpillar is completed in about 4 to 5 weeks. This insect prefers to attack smaller plants with bulbous petioles; petioles that are attacked often become waterlogged and die. However, the impact of the *Niphograpta* caterpillar has been difficult to evaluate because it causes tremendous damage for only a brief period and then disappears. The *Neochetina* weevils and the *Niphograpta* stem-boring caterpillar are established and occur almost everywhere waterhyacinth is distributed throughout the southern US. Growth of waterhyacinth is suppressed and vegetative reproduction is reduced, but other means of control are necessary in most areas.

Hydrilla (Chapter 13.1)
(see http://el.erdc.usace.army.mil/aqua/APIS/apishelp.htm for more information)

Two *Bagous* weevils (one from India that attacks tubers and one from Australia that mines stems) have been introduced as biocontrol agents for hydrilla, but both have failed to establish. However, two *Hydrellia* flies (one from India and one from Australia) have become established. The fly *H. pakistanae* is widespread in the southern US, whereas *H. balciunasi* is localized in distribution. Populations of *Hydrellia* flies have not reached densities high enough to control hydrilla, possibly due to parasitism of the pupae by a native wasp or perhaps other environmental factors. The entire life cycle for both flies is completed in about 3 to 4 weeks, which should allow development of high insect populations. The adventive *Paraponyx* moth from Asia probably entered the US via the aquarium trade and was

Enemy	Type	Origin (Date)	Success	Comments
Bagous affinis	Weevil	India (1987)	Not Established	
Bagous hydrillae	Weevil	Australia (1991)	Not Established	
Cricotopus lebetis	Midge	Unknown (adventive)	Negligible?	Damages growing tips of hydrilla Also see http://edis.ifas.ufl.edu/IN211
Hydrellia balciunasi	Fly	Australia (1989)	Negligible	Found primarily in Texas
Hydrellia pakistanae	Fly	India (1987)	Negligible?	Widely distributed on dioecious hydrilla in the southeastern and south-central US
Paraponyx diminutalis	Moth	Asia (adventive)	Negligible	Causes localized occasional heavy damage to hydrilla
Ctenopharyngodon idella	Fish	China (1963)	Substantial	Throughout the US by permit (Chapter 9) Also see http://plants.ifas.ufl.edu/guide/grasscarp.html

discovered in Florida feeding on hydrilla in 1976. The life cycle of *Paraponyx* is completed in 4 to 5 weeks; the moth was never studied or approved for release, but large populations of hydrilla are occasionally completely defoliated by the moth. The adventive naturalized nonnative *Cricotopus* midge has been associated with hydrilla declines in several Florida locations since 1992. The life cycle of *Cricotopus* is completed in 1 to 2 weeks and developing larvae of the midge mine the shoot tips of hydrilla, which severely injures or kills the plant's growing tips. Feeding damage changes the plant's structure or architecture by preventing new hydrilla stems from reaching the surface of the water column. Despite localized and occasionally severe impacts on hydrilla, none of these insects can cause damage significant enough to provide adequate control when used alone. Research to identify biocontrol agents for hydrilla continues due to the increasing spread of the species throughout the US, its development of resistance to the herbicide fluridone and the relatively high costs associated with other methods employed to control this weed.

Purple loosestrife (Chapter 13.6)
(see http://el.erdc.usace.army.mil/aqua/APIS/apishelp.htm for more information)

Enemy	Type	Origin (Date)	Success	Comments
Galerucella calmariensis	Beetle	Germany (1992)	Substantial	Widely distributed throughout the range of purple loosestrife in the US
Galerucella pusilla	Beetle	Germany (1992)	Substantial	
Hylobius transversovittatus	Weevil	Germany (1992)	Substantial	
Nanophyes marmoratus	Weevil	France, Germany (1994)	Negligible?	

Two nearly identical *Galerucella* leaf beetles are responsible for most biocontrol of purple loosestrife; in fact, these beetles have reduced purple loosestrife infestations by 90% in several states, especially Oregon and Washington. Larvae feed on buds, leaves and stems of the plants and heavily defoliated plants are often killed by the feeding insects. The life cycle of the beetles is completed in about 6 weeks but there is only one generation per year, with pupation occurring in the soil if it is not continuously flooded. This low rate of reproduction is responsible for the lag time between introduction of the beetles and noticeable effects on the plants. Two weevils – the root-attacking *Hylobius* and seed-attacking *Nanophyes* – also contribute to the successful biocontrol of purple loosestrife. Larvae of *Hylobius* feed and develop in the tap roots and pupation occurs in the upper part of the root. Larvae require 1 to 2 years to complete their development and adults can live for several years. Adults of *Nanophyes* feed on young leaves or flowers and lay their eggs in flower buds. Pupation occurs inside the bud and larvae consume the flower buds; buds then fail to open and drop prematurely from the plant. Although the entire life cycle is completed in about 1 month, there is only 1 generation per year. Leaf-eating *Galerucella* beetles, root-attacking *Hylobius* weevils and seed-attacking *Nanophyes* weevils have only recently been introduced as biocontrol agents on purple loosestrife but appear to be very successful in reducing the growth, occurrence and competitiveness of this emergent weed.

Eurasian watermilfoil (Chapter 13.2)

(see http://www.invasive.org/eastern/biocontrol/6EurasianMilfoil.html for more information)

Enemy	Type	Origin (Date)	Success	Comments
Acentria ephemerella	Moth	Europe (adventive)	Negligible?	All can cause declines to populations of Eurasian watermilfoil in localized areas of lakes. Results are difficult to predict.
Cricotopus myriophylli	Midge	China (adventive)	Negligible?	
Eurychiopsis lecontei	Weevil	US (native)	Substantial?	

Several insects have been found attacking Eurasian watermilfoil during overseas surveys, but none have been introduced to the US thus far. Recent declines in the abundance of Eurasian watermilfoil in some northern lakes have been attributed to the adventive *Acentria* moth and *Cricotopus* midge, as well as the native *Eurychiopsis* weevil. These insects are widely distributed throughout the range of Eurasian watermilfoil in North America and are found in all areas infested by the weed; as a result, it is difficult to assess their effectiveness as biocontrol agents. Larvae of the *Acentria* moth feed both in and on stems and leaves, which causes the leaves to drop off the plant. Females have reduced wings and are usually flightless and mating occurs in or on the water surface. Two generations are produced annually and pupae form on the stems. Larvae also feed on a variety of native plants in the absence of Eurasian watermilfoil, so the *Acentria* moth is not a typical biocontrol agent. The *Cricotopus* midge is widely distributed and has been shown to reduce the growth and biomass of Eurasian watermilfoil in laboratory experiments. This midge is not the same species of *Cricotopus* that attacks hydrilla, which suggests these insects may be host specific. It is worth noting that midges rarely feed on living plant tissue and most species typically feed on decaying organic matter. The *Eurychiopsis* weevil is generally considered to be the most important biocontrol agent of Eurasian watermilfoil from an operational perspective even though it is a native insect because this weevil prefers Eurasian watermilfoil over its native natural host. The life cycle of the weevil is completed in about 30 days; adults feed on leaves and stems, whereas larvae are stem borers that consume apical meristems. Feeding damage causes the stems to break apart and heavy feeding by the insects prevents the formation of surface mats. High populations of the *Eurychiopsis* weevil have been associated with declines of populations of Eurasian watermilfoil in some northeastern and midwestern states but fish predation may prevent this weevil from reaching its full biocontrol potential. The *Eurychiopsis* weevil is commercially available and can be purchased to augment existing weevil populations. However, research studying the value of augmenting existing populations with purchased insects has been inconclusive.

Waterlettuce (*Pistia stratiotes*)

(see http://el.erdc.usace.army.mil/aqua/APIS/apishelp.htm for more information)

Waterlettuce is a tropical species that is believed to be native to North America and was extirpated (died out) during the Ice Ages, but was reintroduced into Florida in the 16th century. It forms large floating mats similar to those of waterhyacinth in the extreme southern US and populations of waterlettuce often increase as waterhyacinth populations decline. Waterlettuce is a public health issue in Florida, where larvae of disease-causing *Mansonia* mosquitoes (Chapter 5) attach to the extensive feathery roots to obtain oxygen. Two insects have been released as biocontrol agents of waterlettuce but only the *Neohydronomus* weevil has become established.

Enemy	Type	Origin (Date)	Success	Comments
Spodoptera pectinicornis	Moth	Thailand (1990)	Not Established	May be affected by predation by other insects
Neohydronomus affinis	Weevil	Brazil (1987)	Negligible?	

Adults and larvae of the *Neohydronomus* weevil feed on the leaves, crown and newly emerging shoots of waterlettuce and the characteristic "shot hole" appearance of leaves indicates high weevil densities. Feeding by multiple larvae destroys the spongy leaf bases, which causes plants to lose buoyancy. The life cycle of the *Neohydronomus* weevil is completed in 3 to 4 weeks. The weevil has not contributed to long-term suppression of the plant in the US, but has provided successful biocontrol of waterlettuce in other countries. It is thought that the *Neohydronomus* weevil is heavily preyed upon by imported fire ants in Florida; if true, this provides an interesting example of an exotic invader controlling a valuable potential biocontrol agent.

Giant salvinia (Chapter 13.4)

(see http://el.erdc.usace.army.mil/aqua/APIS/apishelp.htm for more information)

Enemy	Type	Origin (Date)	Success	Comments
Cyrtobagous salviniae	Weevil	Brazil? (adventive)	Negligible, Substantial	Provides good control of common salvinia in FL but not elsewhere. Effects of 2001 introduction on giant salvinia are still being evaluated
Cyrtobagous salviniae	Weevil	Brazil (2001)	Substantial?	

The *Cyrtobagous* weevil is the only insect that has been released as a biocontrol agent of giant salvinia. Adventive weevils that were discovered in Florida in 1960 are used to control common salvinia (*Salvinia minima*), whereas weevils released in 2001 from a Brazilian population are used as biocontrol agents for giant salvinia. The entire life cycle of the *Cyrtobagous* weevil takes about 46 days. Adults feed on leaf buds and leaves and larvae tunnel inside the plant, killing leaves and rhizomes. Attacked plants turn brown and eventually lose buoyancy. *Cyrtobagous* weevils from Australia are currently of great interest to researchers and have been introduced as biocontrol agents for giant salvinia, but it is too early to determine the effectiveness of these weevils in the US.

Melaleuca (*Melaleuca quinquenerva*)

Enemy	Type	Origin (Date)	Success	Comments
Oxyops vitiosa	Weevil	Australia (1997)	Substantial	Not established in permanently flooded sites due to inability to complete life cycle.
Boreioglycaspis melaleucae	Psyllid	Australia (2002)	Substantial	Also see http://edis.ifas.ufl.edu/document_in172
Fergusonina turneri	Fly	Australia (2005)	Not Established	
Lophodiplosis trifida	Fly	Australia (2008)	Negligible	Establishment confirmed http://tame.ifas.ufl.edu/photo_gallery/biocontrol/stem-gall-fly.shtml

Melaleuca is a locally invasive plant that occurs only in south Florida and the Everglades and was introduced multiple times during the early 1900s. The species was used as an ornamental tree and was planted in marshes to drain wetlands. Melaleuca typically grows in dense, impenetrable stands and can attain a height over 50 feet. Four insects have been released as biocontrol agents of melaleuca but only three have become established.

The *Oxyops* weevil and the *Boreioglycaspis* psyllid were released in 1997 and 2002, respectively, and are widely established on melaleuca in south Florida. Damage to the tree is caused primarily by the immature stages of these insects. The slug-like weevil larvae feed on newly expanding leaves; psyllid nymphs attack older leaves and woody stems in addition to new leaves and the psyllid can kill newly emerged seedlings as well. These two insects complement each other well; the psyllid is able to complete its development entirely in the tree canopy under flooded conditions that prevent establishment of the weevil, which must pupate in the soil. Extensive leaf damage from both insects causes melaleuca to divert resources to the production of new foliage instead of flowers. The life cycle of the weevil is completed in about 3 months, whereas a new psyllid generation is produced in 6 weeks. The *Oxyops* weevil and the *Boreioglycaspis* psyllid have contributed to the substantial biocontrol of melaleuca. The *Lophodiplosis* gall-forming fly was released in 2008 and has apparently become established; however, it is too early to assess its impact on melaleuca.

Summary
The use of insects as biological control agents for aquatic weeds has yielded mixed results, which is typical and expected of biocontrol programs. A number of aquatic weeds – including alligatorweed, purple loosestrife and melaleuca – are being successfully controlled by insects released as biocontrol agents for these species. Control of other aquatic weeds – including waterhyacinth, hydrilla, Eurasian watermilfoil, waterlettuce and giant salvinia – has been less successful. Multiple factors play a role in the failure of some biocontrol agents to reach their full potential. For example, the *Neohydronomus* weevil has provided successful biocontrol of waterlettuce in other countries, but has failed to control waterlettuce in Florida, possibly due to predation of the weevil by imported fire ants. Biocontrol can be an effective tool in the aquatic weed manager's arsenal since host-specific biocontrol agents allow management of populations of weedy species while leaving nontarget native plants unharmed. Therefore, it is important that researchers continue to identify and evaluate biocontrol agents so that the successes realized in the control of alligatorweed, purple loosestrife and melaleuca can be duplicated in other weedy aquatic species. A major factor that limits the utility of biocontrol is that unless a potential biocontrol agent is species-specific, it cannot be introduced into the US. Therefore, it is unlikely that biocontrol alone can control all the invasive aquatic weeds in the US.

For more information:
•Harley KLS and IW Forno. 1992. Biological control of weeds: a handbook for practitioners and students. Inkata Press, Melbourne, Australia.

Photo and illustration credits:
Page 55: Alligatorweed flea beetle; Gary Buckingham, USDA-ARS
Page 56: Graph; from Harley and Forno, 1992

Chapter 10

Douglas Colle
University of Florida, Gainesville FL
dcolle@ufl.edu

Grass Carp for Biocontrol of Aquatic Weeds

Introduction

The grass carp or white amur is native to the large river systems of Eastern Asia (China, Siberia) and has been distributed worldwide for use as a food fish and for biological control of aquatic weeds. Natural reproduction of this fish is limited on a world-wide basis due to river modification and reservoir construction, but grass carp are easily produced in aquaculture using artificial means. The fish is a member of the large minnow or Cyprinid family, which includes other fish such as common carp, goldfish and our native minnows and shiners. The grass carp is very different from the well-known common carp, which is also nonnative. Several adaptations equip the grass carp for feeding on plants. For example, the mouth of the grass carp is located high on the head, whereas the mouth on the common carp is positioned low on the head to facilitate bottom feeding in shallow water, which increases turbidity. Also, the grass carp has specialized grinding teeth, which allows it to feed on aquatic plants. Juvenile grass carp consume small invertebrates but become strict vegetarians once they grow to greater than two inches in length. Grass carp are long-lived freshwater fish that can survive for up to twenty-five years if adequate food is available and can grow as much as ten pounds per year. An Arkansas angler caught a grass carp in 2004 that was 53 inches long and weighed 80 pounds. Grass carp can tolerate salinities up to 10 parts per thousand (about 1/3 the salinity of seawater), which allows the species to move through or live in the brackish waters of coastal marshes and estuaries.

History in the US

Grass carp were originally imported into the US in 1963 through a cooperative effort between Auburn University and the United States Fish and Wildlife Service. The species was imported to the US to be evaluated for its potential as a biological control agent for aquatic weeds. Grass carp have been so effective at aquatic weed control that they are now used in 35 different states, primarily for weed control in aquaculture and in closed public or private water bodies. Grass carp that escaped from early stocking programs have formed naturally reproducing populations in the Trinity River system in Texas and throughout the entire Mississippi river drainage system. Most states currently require that only artificially produced sterile triploid grass carp be stocked to prevent further natural reproduction in our remaining river systems. Triploid grass carp are created by shocking fertilized grass carp eggs with cold, heat or pressure, which renders individuals sterile and eliminates any possibility of reproduction. The use of grass carp for aquatic weed control is governed by individual states; some require permits, site inspections and use of sterile fish, whereas others have no restrictions. Many states in the northern US actually prohibit the possession, sale or transportation of grass carp. As a result, you must consult the appropriate state agencies before considering grass carp for weed control to determine whether their use is restricted or prohibited in your state.

Consumption rates and aquatic plant preferences

Grass carp consumption rates (measured as the daily percentage of body weight eaten) are affected by size of the fish and by environmental characteristics such as temperature, salinity and oxygen content of the water. Also, grass carp consumption rates decrease as fish become larger and reach sexual maturity (which occurs even in sterile fish) at 2 or 3 years of age. Large grass carp (over 15 pounds) consume up to 30% of their body weight daily, whereas smaller fish (less than 10 pounds) can consume as much as 150% of their body weight a day. Maximum consumption occurs when water temperatures range from 78 and 90 °F and is greatly reduced at temperatures below 55 degrees. Consumption is reduced by 45% when oxygen levels in the water drop to 4 ppm and fish stop feeding completely if the oxygen level drops below 2 ppm. Although grass carp can tolerate salinities up to 10 parts per thousand, they will not feed if salinity levels are higher than 6 parts per thousand.

Grass carp are general herbivores and will eat almost any plant material, including grass clippings, young waterlilies and even cattail shoots. The species does, however, have preferences for some plants, including southern naiad, hydrilla (Chapter 13.1) and duckweed. Although grass carp do show preferences for certain plant species, they are vegetarians and will consume almost all other submersed aquatic vegetation once populations of their preferred species have been depleted. Eurasian watermilfoil (Chapter 13.2) is, however, an exception to this rule. Grass carp stocked in Deerpoint Reservoir in Florida have controlled all the hydrilla in the reservoir, but populations of Eurasian watermilfoil have increased following hydrilla removal. Grass carp are poor biocontrol agents of filamentous algae (Chapter 12), spatterdock, fragrant waterlily, sawgrass, cattail and other large plants.

Variables that affect stocking rates and duration of aquatic plant elimination

Grass carp should not be stocked in open systems that are connected to a stream or river because they migrate with moving water and will leave the stocked water body. Grass carp stocking rates in closed systems typically range from 2 to 50 fish per acre; the price of triploid grass carp ranges from 10 to $20 per fish and is dependent on proximity to the producer, the distance the fish must be transported and the size of fish desired. Most biologists agree that there is no "magic number" of grass carp to stock to achieve a specific percentage of submersed weed control because optimum stocking rate is dependent upon the type and quantity of aquatic plants present, water temperature, oxygen content and desired speed of weed control. Once grass carp are stocked, predation by fish-eating predators can be a problem because grass carp typically feed near the water surface and are commonly preyed upon by osprey, otters and other fish. For example, studies in research ponds in Florida revealed that the number of grass carp lost to predation ranged from 7 to 70% one year after stocking. Predation can be especially problematic in water bodies with large fish predators such as striped bass or largemouth bass. Grass carp that are larger than 12 inches should be used in these systems to avoid losing the majority of the stocked grass carp to predation and to ensure adequate aquatic weed control. Overstocking or excellent survival of grass carp results in removal of almost all submersed aquatic plants, whereas understocking or excessive mortality of grass carp results in no noticeable plant control. The proper balance of grass carp and weed growth is difficult to achieve and varies among waterbodies.

Complete elimination of aquatic plants by grass carp can be maintained for as long as fifteen years in the southern and the southwestern US if enough fish are initially stocked to consume the aquatic vegetation in the system, whereas control can last for up to 10 years in the rest of the country. It is important to remember that the use of grass carp as biocontrol agents is a long-term strategy because grass carp grow to an extremely large size, live up to 25 years and cannot easily be removed from a water body once they are stocked. In fact, it is not possible to remove significant numbers of grass carp from large lakes in a timely fashion. For example, significant numbers of grass carp have been removed by bow fishermen in Caney Lake in Louisiana but only after several years of effort.

Effects on water quality and fish populations
Total elimination of aquatic vegetation by grass carp usually results in changes in water quality because the water body shifts from a plant-based community to a system dominated by phytoplankton and/or algae (Chapter 1). Long-term increases in chlorophyll, total phosphorus and nitrogen often accompany the shift to a phytoplankton-based system once grass carp consume all the aquatic vegetation. In addition, water clarity usually decreases due to the increase in algae and/or phytoplankton and to wind and wave action that stirs up and suspends bare sediments.

Populations of fish species that require aquatic plants for spawning, nursery areas or as a feeding source will likely experience rapid declines or may be eliminated from the system altogether (Chapter 2). Grass carp eliminated all aquatic vegetation for 15 years in two Florida lakes, which caused the total loss of all populations of pickerel, taillight shiner, golden topminnow, bluespotted sunfish and Everglades pygmy sunfish. Both lakes also had large declines in lake chubsucker, golden shiners and warmouth populations. In contrast, tremendous increases were noted in populations of both gizzard shad and threadfin shad after plant removal because these species feed on phytoplankton. Largemouth bass, bluegill and redear sunfish populations were not affected by elimination of all aquatic plants during this time period.

Summary
Grass carp can be an effective, cost-efficient tool for long-term aquatic plant removal in closed systems. One of the initial concerns regarding the use of this fish as a biocontrol agent in the US was the potential of escaped fish to reproduce in the wild because diploid fish have escaped into several

river systems and natural reproduction has been documented in the Mississippi and Trinity River watersheds. However, the development and aquaculture production of sterile triploid grass carp has provided a solution to this problem. Grass carp remain illegal in many states and most other states require permits for use of the fish. In states where their use is allowed, utilization of grass carp as a biocontrol agent for aquatic weed management can be a very effective strategy provided decade-long control is desired and users accept that there is no way to efficiently remove the fish once they are stocked in the system.

For more information:
- Aquatic plant management – triploid grass carp. Washington State Department of Ecology. http://www.ecy.wa.gov/programs/wq/plants/management/aqua024.html
- Brunson M. 2007. Grass carp in Mississippi farm ponds. Publication 1894. Extension Service of Mississippi State University. http://msucares.com/pubs/publications/p1894.htm
- Grass carp control: weeds in ponds and lakes. 1999. Missouri Department of Conservation Aquaguide. http://mdc4.mdc.mo.gov/Documents/15.pdf
- Masser M. 2002. Using grass carp in aquaculture and private impoundments. Publication No. 3600. Southern Regional Aquaculture Center. http://www.aquanic.org/publicat/usda_rac/efs/srac/3600fs.pdf
- Proceedings of the grass carp symposium. 1994, 1979. Gainesville, Florida. http://plants.ifas.ufl.edu/guide/grasscarp_proceedings.html
- Sutton DL and VV Vandiver, Jr. 2006. Grass carp: A fish for biological management of hydrilla and other aquatic weeds in Florida. Bulletin 867. Florida Cooperative Extension Service of the University of Florida. http://edis.ifas.ufl.edu/FA043
- Triploid grass carp information. 2008. Texas Parks and Wildlife Department. Inland Fisheries Division. Austin, Texas. http://www.tpwd.state.tx.us/landwater/water/habitats/private_water/gcarp.phtml
- Triploid grass carp in New York ponds. New York Department of Environmental Conservation. http://www.dec.ny.gov/outdoor/7973.html

Photo and illustration credits:
Page 61: Mature grass carp; William Haller, University of Florida Center for Aquatic and Invasive Plants
Page 62: Releasing grass carp; William Haller, University of Florida Center for Aquatic and Invasive Plants
Page 63: Grass carp; Paul Shafland, Florida Fish and Wildlife Conservation Commission

Chapter 11

Michael D. Netherland
US Army Engineer Research and Development Center
Gainesville FL
mdnether@ufl.edu

Chemical Control of Aquatic Weeds

Introduction

Chemical control through use of registered aquatic herbicides and algicides is a technique that is widely employed by aquatic plant managers in both private and public water bodies throughout the United States. Treatments can range in size from backpack spray applications to individual plants or small clusters of plants up to large-scale treatments from boats or helicopters that may target an invasive weed throughout an entire lake. In addition, the objective of some treatments is broad-spectrum control of numerous plant species, while many treatments target a specific invasive plant or algal species. The difference in scale, scope, timing, regulations and management objectives associated with the use of aquatic herbicides makes it a challenge to write an all-encompassing single chapter. In this document we seek to explain some of the rules and regulations associated with aquatic herbicide labeling, explain trade, chemical and common names, describe key differences between submersed and emergent applications, contrast contact and systemic herbicides, and provide specific information on each registered aquatic herbicide.

All herbicides discussed in this chapter have undergone EPA review (Appendix A) and have been approved for aquatic use. This does not mean these herbicides are registered or can be used in every state as all states have their own regulatory and registration procedures. In addition, some states require applicators of aquatic herbicides to be certified and licensed before these products can be used. Many states also require that permits be obtained before herbicides can be applied to bodies of water – even if the waters are privately owned. Herbicide labels and MSDS (material safety data sheets) are available online on the registrant's website and are excellent sources of information. _Always read the label on the herbicide and check with the appropriate regulatory agencies in your state before purchasing or applying pesticides to any body of water._

Like all pesticides, aquatic herbicides have three names: a trade name, a common name and a chemical name. The trade name of a product is trademarked and is owned by the company, whereas the common name and the chemical name are assigned by the American National Standards Institute and the rules of organic chemistry, respectively. For example, consider the aquatic herbicide Rodeo®. The trade name of this herbicide is Rodeo®, the common name is glyphosate and the chemical name is N-(phosphono-methyl) glycine, isopropylamine salt. If a particular pesticide is protected by a patent, there may be only a single trade name associated with that pesticide. However, if the pesticide is off-patent, there may be multiple trade names that share single common and chemical names. A number of aquatic herbicides are off-patent and have multiple trade names; therefore, we refer to herbicides by their common names only throughout most of this handbook.

There are approximately 300 herbicides registered in the US, but only around a dozen are registered for use in aquatic systems. Herbicide labels often include a list of the nuisance species controlled by the product, but applicators may be allowed to use the herbicide to control a target weed not listed on the herbicide label provided the product is labeled for use at the desired site of application. For example, if you wish to use an herbicide to control a weed in your pond and the weed is not listed on the herbicide label, you may still be able to use the product to control this particular weed if the label specifies that the herbicide may be used in ponds. However, it is important to check with state authorities before doing so because some states specify that herbicides can only be used to control weeds that are listed on the product label. Additionally, the user accepts liability for the performance of the product if the specific weed is not included on the label.

Herbicides can be classified in several ways, including by their chemical family, their mode of action (how they work) and their time of application in relation to growth of the weed. In this handbook we will classify aquatic herbicides based upon how they are applied (as foliar or submersed treatments – although some herbicides are both) and on their activity in the plant (systemic or contact).

Products that are applied as foliar treatments are most easily recognized by the public. For example, if you have a weed to control, you select a herbicide based on label directions, mix the product with the prescribed amount of water and apply it directly to the weed. Contact products work quickly and kill the pest rapidly on contact (hence the designation "contact"). Systemic compounds, on the other hand, usually work slowly by affecting biochemical pathways and must be absorbed by the plant before providing control; therefore, systemic compounds may require days or weeks to kill the weed. The application method is the same for both systemic and contact herbicides – the compound is applied at a fairly high concentration directly to the foliage of the plant. Foliar herbicides are used to control floating, floating-leaved and emergent aquatic weeds.

Submersed herbicides are applied as concentrated liquids, granules or pellets. Liquid treatments are mixed with water to facilitate application and to ensure even distribution and are applied to the entire water column to control submersed weeds and planktonic algae. Some dry formulations are also mixed with water, but many granular and pelleted products are applied using granular spreaders. Aquatic herbicide applicators must measure the volume of the water to be treated before applying submersed herbicides to ensure that the appropriate and effective amount of herbicide is used. The following constants are needed to calculate the volume of water before treatment with submersed herbicides:

- The volume of a body of water is calculated in acre-feet, which is a function of area and depth; for example, a lake with an area of 1 acre and a depth of 6 feet has a volume of 6 acre-feet
- A single acre-foot of water comprises around 326,000 gallons of water and weighs around 2.7 million pounds

The volume (in acre-feet) of a body of water or treatment site is used to determine the amount of herbicide needed to control a particular weed. For example, if the label of a herbicide specifies an application rate of 1 ppm (part per million), then 2.7 pounds of the herbicide's active ingredient must be applied for each acre-foot of the water to effectively control the target weed. This results in a concentration of 1 ppm since 2.7 pounds of herbicide are mixed with 2.7 million pounds of water in each acre-foot. Most herbicide labels include a table that lists application rates, but it may be necessary to perform calculations similar to those described above to ensure that the correct dosage of herbicide is applied. The labels of aquatic herbicides clearly state how these calculations are performed.

Contact Herbicides

Several herbicides registered for aquatic use are classified as contact herbicides. This term may lead one to believe that these herbicides kill weeds immediately after contacting them. While contact herbicides do tend to result in rapid injury and death of the contacted plant tissues, it is important to realize that the term "contact herbicide" refers to the lack of translocation or mobility of the herbicide in the plant after the herbicide is taken into the plant tissue. Herbicides that are able to move through plant tissues following uptake are said to translocate; these products are called "systemic herbicides." This distinction between contact and systemic herbicides has significant implications for the prescribed use of the products and usually describes how quickly weeds may be controlled.

Contact herbicides are often used for foliar treatment of sensitive free-floating plants such as waterlettuce, duckweed and salvinia (Chapter 13.4) and good spray coverage is essential to ensure control of all individual plants of these species. Contact herbicides are also used to temporarily control emergent aquatic plants. These treatments are often initially effective, but treating large emergent plants with a contact herbicide often results in rapid recovery and significant regrowth from plant tissues that do not come into contact with the herbicide. As a result, systemic products are usually preferred for controlling emergent plants because systemic herbicides move or translocate within the plant and kill underground roots and rhizomes, which reduces or eliminates regrowth.

Contact herbicides that are used to control submersed weeds must remain in the water of the treatment area for a few hours to a few days so that plants are exposed to a lethal concentration of the herbicide for a sufficient amount of time. The results of an herbicide application designed to control submersed plants is primarily impacted by two key factors:

1) the concentration of the herbicide in water that surrounds the target plant
2) the length of time a target plant is exposed to dissipating concentrations of that herbicide

This dose/response phenomenon is herbicide- and plant-specific and has been defined as a concentration and exposure time (CET) relationship. Contact herbicides have relatively short exposure times (often measured in hours or days), which means that these products are used to target specific areas within a larger water body or in areas where significant dilution is expected. Whether for contact or systemic herbicides, the vast majority of poor treatment results following submersed applications are due to an inability to maintain the herbicide in contact with the target plants at an effective concentration for an appropriate period of time. Each contact herbicide has a different use rate, exposure requirement and selectivity spectrum. While the registered contact herbicides are often referred to as "broad-spectrum" products, there is a range of plant susceptibilities to each of these contact herbicides based on the species, use rate, treatment timing and exposure period. Proper identification of target and nontarget plants is important when selecting a contact herbicide because herbicides usually differ significantly in their selectivity to various plant species.

Susceptible submersed plants that are treated with contact herbicides typically show symptoms of herbicide damage within a day or two of treatment and collapse of the target plants can occur within 3 to 14 days. It is important to note that the use of contact herbicides in areas with dense plant populations and warm water temperatures can lead to a situation where decomposing plant tissue quickly depletes the oxygen from the water column, resulting in conditions that can cause a fish kill. Product labels have directions that provide guidance to avoid oxygen depletion when treatments are made under conditions of dense vegetative cover and warmer water temperatures.

It is important that contact herbicides be applied and distributed as evenly as possible to the target plant (or throughout the water column for control of submersed plants) to ensure that the entire plant – including the rooted portions of the plant near the sediment – is exposed to the herbicide. Poor mixing of contact herbicides within the water column can result in control of plant tissue growing near the water surface, followed by rapid recovery from the lower portions of the plant that were not exposed to the herbicide. Poor control can also result from summer applications when treating lakes that are thermally stratified (Appendix B).

Contact herbicides are currently used for both small-scale treatments such as along shorelines and for large-scale control efforts. Most of the contact herbicides have been registered for many decades and they tend to be versatile with a very wide range of use patterns. Combinations of two or more contact herbicides are often used to target specific invasive or nuisance species. The registered contact herbicides (and dates of registration) are described in more detail below. These brief descriptions are not comprehensive, but are meant to serve as a guide to particular historical strengths or potential issues associated with the use of these products.

Compound/ Date registered for aquatic use	Primary use – Submersed	Primary use – Floating	Primary use – Emergent	Mode of action	Comments
Copper 1950s (Liquid, granular)	X	X		Contact Plant cell toxicant	Algae control Also used in combination with other herbicides Often used for submersed plants near potable water intakes
Endothall 1960 (Liquid, granular)	X	X		Contact Inhibits respiration and protein synthesis	Dipotassium salt for submersed plant control Dimethyl-alkylamine salt for algae and plants that are more herbicide-tolerant Treatment timing affects selectivity
Diquat 1962 (Liquid)	X	X	X	Contact Inhibits photosynthesis and destroys cell membranes	Broad spectrum Turbidity affects effectiveness Very fast activity on sensitive plants
Carfentrazone 2004 (Liquid)		X	X	Contact Inhibits plant specific enzyme	Waterlettuce and broadleaf weed control

Copper (1950s)

Copper is a micronutrient that is needed for healthy growth of animals and is often added to animal feed and to vitamins formulated for human use. Copper is widely used as a fungicide in agricultural systems to control diseases on food crops and copper-based compounds have been used for aquatic plant control since the early 1900s. Copper sulfate is likely the most widely used copper product, but it is corrosive and its effectiveness can be affected by water hardness. Liquid chelated copper compounds were developed in the 1970s to reduce these problems. Copper compounds are used primarily to control algae and plants growing in irrigation canals, ponds, lakes and reservoirs. Submersed use rates typically range from 0.2 to 1.0 mg/L copper in the water column. There are no restrictions on the use of copper in potable water sources or in waters used for crop irrigation. This allows the immediate use of treated water and helps to explain why copper is widely used to control nuisance plants in drinking water supplies and irrigation canals. Copper acts very quickly on plants and algae and has a short exposure requirement, which can be advantageous when treating small areas or areas subject to rapid dilution. The effectiveness of copper as a herbicide or algicide can be affected by alkalinity or hardness of the water. For example, high alkalinity or hard water can reduce the effectiveness of copper-based products. Despite these limitations, copper remains the major tool for algae control in potable water systems, irrigation canals and in small water bodies. Copper does not biodegrade and regular use can result in increased copper residues in the sediment. Copper is generally considered to be biologically inactive in sediments.

Diquat (1962)

Diquat is a fast-acting contact herbicide that disrupts photosynthesis in susceptible plant species. Diquat effectively controls many free-floating weeds including duckweed, watermeal, waterlettuce and salvinia. As noted above, good coverage is critical when treating these plants because missing a small area or a few individuals can lead to rapid recolonization by these fast-growing floating species. Diquat is also used to control submersed plants in small treatment areas or in areas where dilution may reduce the period of time that plants are exposed to the herbicide. Diquat is generally considered to be a "broad-spectrum" product that kills a wide range of plant species. However, the susceptibility of different submersed species can vary significantly. Diquat can be rapidly inactivated when treating "muddy" or turbid water and the speed of this inactivation can interfere with plant control. There are no hard and fast rules to determine when water is too muddy to treat, but the effectiveness of diquat increases as water clarity increases. Diquat is often mixed with copper-based herbicides to control a broader range of weeds and to improve control of target plants. In addition to its use in aquatic systems, diquat is labeled for weed control in turf and along fence lines and has been used to kill the leaves and vines of potato to increase ease of harvesting.

Endothall (1960)

Endothall is used primarily to control submersed plants and use rates and methods of application vary significantly. Traditional use patterns of endothall have included spot treatments of small target areas with a granular product. These spot treatments are generally applied at the highest label rate and species selectivity is not a major concern. Selective use of the product is based on species sensitivity, use rates and treatment timing. The effectiveness of endothall is generally not affected by factors such as alkalinity or turbidity of the water. Within the last several years, large-scale early-season treatments have been applied to target invasive plants such as hydrilla (Chapter 13.1), curlyleaf pondweed (Chapter 13.7) and Eurasian watermilfoil (Chapter 13.2) that persist throughout the winter. These treatments are conducted before desirable native plants begin to grow in spring, which may allow control of the invasive weeds with limited impact on native species that grow later in the season. It is important to note that these early-season treatments are applied when plant biomass is not at its peak and when water temperatures are cooler. These conditions reduce or prevent oxygen depletion that may occur when fast-acting contact herbicides are applied to dense nuisance populations of weeds in warmer water. Endothall has also been used to control weeds in turf and to desiccate cotton and potato plants to aid in harvesting.

Carfentrazone (2004)

Carfentrazone is the most recently registered contact herbicide and the major use patterns are still being established. Carfentrazone is the only contact herbicide that impacts a plant-specific enzyme. In contrast to the other registered contact herbicides, carfentrazone has a fairly narrow spectrum of control, which leads to improved selectivity and reduces damage to nontarget plants. However, this limits the use of carfentrazone to a few weed species. To date, carfentrazone has been used for control of waterlettuce and selected broadleaf emergent plants. Carfentrazone is labeled for submersed plant control; however, variable milfoil is currently the only plant that has shown a high level of susceptibility to this herbicide. Carfentrazone is also used for weed control in turf, corn and other crops.

Peroxides (1980s)

Several inorganic chemicals produce peroxide (principally hydrogen peroxide) when mixed with water. This contact herbicide is used in aquatic systems only for control of algae and has no effect on submersed vascular plants. Blowers or granular spreaders are used to ensure uniform coverage of the water surface. These compounds produce hydrogen peroxide—which is toxic to some species of algae—when they come into contact with water. Hydrogen peroxide then rapidly breaks down into water, oxygen and other natural products. Often used to control algae in potable water supplies, hydrogen peroxide is also widely used in the medical field to kill bacteria.

Systemic herbicides – auxin mimics

In contrast to the contact herbicides, systemic herbicides are mobile in plant tissue and move through the plant's water-conducting vessels (xylem) or food-transporting vessels (phloem). Once the herbicide is absorbed into the plant, it can move through one or both of these vessels and throughout the plant tissue to affect all portions of the plant, including underground roots and rhizomes. Auxin mimic herbicides simulate auxin, a naturally occurring plant hormone that regulates plant growth. These herbicides generally target broadleaf plants (dicotyledons or dicots) and are often called "selective herbicides" because many aquatic species (particularly grasses or monocots) are not susceptible to auxin mimic herbicides. In fact, the majority of submersed aquatic plants are monocots, which aids in selectivity when using an auxin mimic. After treatment, the shoot tissue of susceptible plants will often bend and twist (epinasty) and plants will often collapse 2 to 3 weeks after herbicide application. Similar to contact herbicides, auxin mimics that are used to control submersed weeds must remain in the treated area for a few hours to a few days so that plants are exposed to a lethal concentration of the herbicide for a sufficient amount of time. Longer exposure periods (such as 24 to 144 hours) increase the probability that the target weed will be completely controlled, but exposure times of 12 hours or greater may provide good control, provided the application rate and timing are appropriate. The contact herbicides discussed above are used to control a large number of nuisance and invasive plant species, but auxin mimic herbicides are used for a much smaller plant spectrum.

Compound/ Date registered for aquatic use	Primary use Submersed	Primary use Floating	Primary use Emergent	Mode of action	Comments
2,4–D 1959 (granular ester) 1976 (liquid amine) 2009 (granular amine)	X	X	X	Systemic Auxin mimic, plant growth regulator	Used for submersed dicots such as Eurasian watermilfoil and for waterhyacinth management
Triclopyr 2002 (liquid amine) 2006 (granular amine)	X	X	X	Systemic Auxin mimic, plant growth regulator	Used for submersed dicots such as Eurasian watermilfoil; also for floating and emergent plants

While there are several aquatic dicotyledons (and some monocots) that show sensitivity to the auxin mimics, these herbicides have historically been used for selective control of a limited number of emergent, floating and submersed plants, including waterhyacinth (Chapter 13.5) and Eurasian watermilfoil. The auxin mimics 2,4–D and triclopyr have very similar use patterns and are used to control broadleaf plants growing among desirable grasses or native submersed plants. This is referred to as "selective control" and is very important in aquatic sites to maintain native species while reducing growth of invasive weeds. These herbicides are also used to control weeds in turf, pastures, forestry and other terrestrial sites.

2,4–D (1959)
A number of nuisance emergent and submersed plants are controlled by 2,4–D, but this herbicide is primarily used for selective control of waterhyacinth and Eurasian watermilfoil. A liquid amine formulation is used to control emergent and submersed plants and a granular ester formulation is used for submersed weed control. In addition, a granular amine formulation has been recently registered. Some native emergent plants – including waterlilies, spatterdock and bulrush – are susceptible to 2,4–D, so care should be taken to avoid injury to these plants. 2,4–D has been used for more than 50 years to control broadleaf weeds in pastures, crops, turf and aquatics.

Triclopyr (2002)
Triclopyr was registered for aquatic use in 2002 and to date the major use of his herbicide has been for selective control of Eurasian watermilfoil. Similar to 2,4–D, there are certainly other plant species that are susceptible to triclopyr; however, the historical strength of auxin mimic herbicides has been selective control of invasives such as Eurasian watermilfoil or waterhyacinth. Triclopyr is registered as both liquid and granular amine formulations. Like 2,4–D, a number of native emergent plants are susceptible to triclopyr, so care should be taken to avoid injury to these plants. The use of triclopyr in public waters is permitted in some states where 2,4–D use is not allowed. Triclopyr is also labeled for control of broadleaf weeds in turf, forestry and crop production.

Systemic herbicides – enzyme specific herbicides for foliar use
Two aquatic herbicides – glyphosate and imazapyr – are labeled only for foliar treatment and control of emergent and floating plants. Both are systemic and readily move through plant tissue to control aboveground and underground portions of the plant. These herbicides inhibit enzymes that plants need to produce proteins that are required for growth, so plants treated with these systemic herbicides slowly "starve" and eventually die. These herbicides target enzymes in a pathway that is found only in

Compound/ Date registered for aquatic use	Submersed	Floating	Emergent	Mode of action	Comments
Glyphosate 1977 (Liquid)		X	X	Systemic Inhibits plant specific enzyme, EPSP inhibitor	Broad spectrum Plant death may be slow Not active in soil – cannot be used in irrigation ditches
Imazapyr 2003 (Liquid)			X	Systemic Inhibits plants specific enzyme (ALS-inhibitor)	Broad-spectrum for emergent plant control

plants. Herbicides that target plant specific enzymes typically show very low toxicity to non-plant organisms such as mammals, fish and invertebrates. Both of these herbicides are truly broad-spectrum and a very limited number of emergent plant species can tolerate exposure to them. These herbicides are especially effective at controlling large monotypic stands of nuisance emergent plants such as phragmites (Chapter 13.9), cattail and other invasive perennial plants that have extensive rhizome and root systems. Both products result in fairly slow control of target weeds and are often mixed together for plants that are particularly hard to control.

Glyphosate (1977)

Glyphosate is widely used in agriculture, homeowner and specialty markets, including aquatics, where glyphosate has been used for control of aquatic weeds since the mid-1970s. Glyphosate is translocated through treated plant tissues; new growth is disrupted and plants die 1 to 4 weeks after herbicide application. Glyphosate has no soil activity and is rapidly deactivated in natural waters and therefore cannot be used for control of submersed weeds. Because this herbicide is rendered inactive so quickly, the irrigation or potable water restrictions associated with the use of glyphosate are minimal. Treatment timing can impact the effectiveness of glyphosate and nuisance species should be treated during late summer or fall when plants are rapidly moving sugars to storage organs such as roots or rhizomes in preparation for overwintering. This treatment strategy can increase the translocation of glyphosate into the storage organs and can result in enhanced control of the target plant during the following growing season.

Imazapyr (2003)

Imazapyr is also used in forestry and specialty markets, including aquatics, where imazapyr was registered for control of aquatic weeds in 2003. Imazapyr inhibits the plant specific enzyme acetolactate synthase (ALS), which plays a critical role in protein production in plants. This herbicide has been used to control invasive plants such as spartina or phragmites that have invaded previously unvegetated areas in tidal zones or river flats. As with glyphosate, imazapyr readily translocates throughout plant tissue and new growth is inhibited due to the lack of protein production. Imazapyr should be applied when the plants are actively growing in the spring, summer or fall and is absorbed through plant leaves and roots. Unlike glyphosate, imazapyr is active in the soil so care should be taken to avoid treating areas around the root zones of desirable plants.

Systemic herbicides – in water use only
Fluridone (1986)

Fluridone is a bleaching herbicide that targets a plant specific enzyme that protects chlorophyll, the green pigment responsible for photosynthesis in plants. Fluridone is the only herbicide registered by the EPA that is labeled only for use in aquatic systems and it is used primarily to control submersed (e.g., Eurasian watermilfoil, hydrilla, and egeria) and floating plants (e.g., duckweed, watermeal and salvinia) by treating the water column instead of the foliage of the plants. Fluridone symptoms are unique and highly visible, with the new growth of sensitive plants bleaching or turning white as chlorophyll in the plant is destroyed by sunlight. Susceptible plants will show bleaching symptoms in new shoot growth; however, it is important to note that bleaching symptoms don't always equal control and actual plant death may not occur for months after an initial treatment. Fluridone has been described as both a selective and broad-spectrum herbicide because use rates can vary from 4 to 150 ug/L. Higher rates often provide broad-spectrum control, whereas lower rates effectively control only a few species.

Unlike the contact or auxin mimic herbicides that require hours or days of exposure, the fluridone label states that target weeds must be exposed to fluridone for a minimum of 45 days. Required exposure periods will often depend on the plant species, stage of plant growth and treatment timing. During the exposure period, new shoot growth of susceptible plants becomes bleached and this continuous bleaching of new growth depletes the plant's reserves of carbohydrates needed for growth. This slow death (which may take 2 or more months) can be beneficial to the environment because plants continue to provide structure for habitat and produce oxygen through photosynthesis. The inhibition of weed growth can also allow native plants to regrow if they are naturally tolerant of fluridone, but regrowth is highly dependent on herbicide rate. The extended exposure requirement typically calls for treatment of the entire aquatic system or treatment of protected embayments of lakes or reservoirs. Despite the extended herbicide exposure requirements associated with fluridone treatments, there are no restrictions for potable water use, fishing or swimming; however, irrigation restrictions are described on the product label. The ability to apply low use rates in the part per billion range, extended exposure requirements and slow plant death have allowed fluridone to be used for numerous whole-lake management treatments throughout the United States targeting invasive plants such as hydrilla and Eurasian watermilfoil.

Fluridone is available in both liquid and pellet formulations. Both products require that plants be exposed to sufficient concentrations of fluridone for an appropriate period of time. As a result, sequential fluridone treatments – often called "bumps" – are usually applied over a period of time to ensure that an effective concentration of the herbicide is maintained. A commercial assay that measures fluridone residue levels is available through the manufacturers of fluridone and can be used to identify current concentrations of fluridone to determine if further applications are necessary to maintain an effective concentration of the herbicide.

Fluridone is also applied to the water column to control floating plants such as duckweed, salvinia and watermeal in small water bodies. Floating plants are generally controlled much more quickly than submersed species. Fluridone is very flexible and can be used in systems of less than one acre and in systems that exceed several thousand acres. Regardless of the size of the treatment, target plants must be exposed to sufficient concentrations of fluridone for an appropriate period of time in order to effectively control the target plant.

Compound/ Date registered for aquatic use	Primary use Submersed	Primary use Floating	Primary use Emergent	Mode of action	Comments
Fluridone 1986 (Liquid, granular)	X	X		Systemic Inhibits plant specific enzyme New growth bleached	Large-scale or whole-lake management Low use rates, long exposure requirements Treatment timing and use rate affect selectivity Used for some floating plants

Systemic herbicides – ALS herbicides
The most recently registered herbicides include two compounds that target the plant specific enzyme acetolactate synthase (ALS). As noted above for imazapyr, this enzyme plays a key role in the production of amino acids needed for protein synthesis in plants and the affected pathway does not occur in animals. In contrast to the broad-spectrum activity described for glyphosate and imazapyr above, the newly registered ALS herbicides tend to be fairly selective. Despite a similar mode of action, the use patterns can vary substantially between these products. Similar to other enzyme specific inhibitors, these herbicides result in a slow kill of the target weed and susceptible floating plants are often controlled much more quickly than large emergent rooted plants. Control of submersed plants tends to occur over a long period of time; although systemic ALS herbicides do not result in bleaching

of new plant growth, they are similar to fluridone in the length of time required to achieve control of submersed plants.

Compound/ Date registered for aquatic use	Primary use Submersed	Primary use Floating	Primary use Emergent	Mode of action	Comments
Penoxsulam 2007 (Liquid)	X	X		Systemic Inhibits plant specific enzyme (ALS) New growth stunted	Large-scale control of hydrilla Floating plant control Extended exposure required for submersed plant control
Imazamox 2008 (Liquid)	X	X	X	Systemic Inhibits plant specific enzyme (ALS) New growth stunted	Selective emergent plant control Waterhyacinth control Growth regulation of hydrilla

Penoxsulam (2007)

Penoxsulam was registered for aquatic use in 2007 and is currently used to control floating species such as waterhyacinth, waterlettuce and salvinia and submersed plants such as hydrilla. Treatments may include foliar application of penoxsulam directly to the target floating plants or submersed application for control of both submersed and floating plants. For submersed applications, penoxsulam exposure requirements are generally similar to those of fluridone and plant death may occur over a period of 60 to 120 days or longer depending on the plant species, stage of plant growth and treatment timing. During the exposure period, new shoot growth is inhibited and plants will often turn red in color. Slow plant death reduces the chance of oxygen depletion in the water and the extended exposure period requirement typically requires treatment of the entire aquatic system or treatment of protected embayments of lakes or reservoirs where dilution from water exchange is minimized. Despite the extended herbicide contact time associated with penoxsulam treatments, there are no restrictions on use of water for drinking, fishing or swimming, but irrigation restrictions are described on the product label. Penoxsulam is also registered for weed control in rice and turf; because it was recently registered for aquatic use, optimum aquatic use patterns are still under development.

Imazamox (2008)

Imazamox was registered for aquatic use in 2008 and is currently used for selective control of large emergent species such as phragmites, Chinese tallow and wild taro, floating species such as waterhyacinth and submersed plants such as hydrilla. Emergent and floating plant use patterns are very similar to imazapyr; however, imazamox is often used in situations where greater selectivity is desired. Use of imazamox for submersed plant control has generally focused on hydrilla and treatments may provide biomass reduction and extended growth suppression (3 to 7 months) of the weed, even when exposure periods are much shorter than those noted for other enzyme inhibitors. Imazamox is registered for weed control in turf and rice, but the most effective use patterns for imazamox in aquatic systems are still being determined because the herbicide was only recently registered for use in water. There are no restrictions on the use of the imazamox-treated water for drinking, fishing, swimming and minimal restrictions for irrigation.

Herbicide resistance and resistance management

Aquatic plant management was thought to be largely unaffected by issues related to herbicide resistance. Nonetheless, the discovery of large-scale resistance of formerly sensitive populations of hydrilla to the herbicide fluridone in Florida during 2000 and 2001 was an unexpected development that has made aquatic managers much more sensitive to this issue. The biochemical basis for resistance development is beyond the scope of this document; however, factors that are known to foster development of resistance include: 1) repeated use of the same herbicide within and over multiple seasons; and 2) use of herbicides that target plant specific enzymes (e.g. ALS inhibitors). When possible, managers should consider rotation of herbicides to reduce the potential for resistance development. In addition, if a manager observes a formerly sensitive target plant population showing a significant change in response to a herbicide, they should immediately contact an aquatic weed specialist for further evaluation of the situation.

Herbicide dissipation and half-lives

The length of time a herbicide remains in contact with target plants following a submersed application is critical to achieving desired results. The two key processes that dictate the required exposure of plants to herbicides are herbicide dispersion and degradation. Once applied to the water, herbicides are subject to dispersion or movement both within and away from the treated area. Dispersion initially has a positive influence on the treatment as it facilitates mixing of the herbicide in the water column. The rate of movement of herbicide residues from the treatment area is likely the largest single factor affecting treatment success, especially for those treatments applied to a small area in a larger water body. For example, application of a herbicide to a 10-acre protected cove in a large reservoir may result in limited movement outside of the treatment area and a subsequent long exposure period. In contrast, a 10-acre plot applied along an unprotected shoreline of the same reservoir on the same day may result in the herbicide moving out of the target area and becoming greatly diluted within a few hours of treatment. Conditions on the day of treatment can also be very important, especially for treatments applied to unprotected areas of larger lakes. High winds or high water flow associated with recent precipitation can have a strong negative influence on treatment results. Because the potential range of exposure periods can vary significantly at the same site from day to day, even greater variation between sites is likely. This variation in the expected exposure period will often influence both choice and application rate of the selected herbicide.

In addition to dispersion, herbicide degradation plays a significant role in the effectiveness of a treatment. With the exception of copper (a natural element), all herbicides are subject to degradation pathways that ultimately lead to breakdown products that include carbon, hydrogen and other simple compounds. These degradation pathways result in decomposition of the herbicide to simpler products that lack herbicidal activity via processes such as photolysis (breakdown by ultraviolet rays in sunlight), microbial degradation (breakdown via action of the microbial community) or hydrolysis (breakdown via the action of water splitting the herbicide molecule). Environmental conditions such as temperature, hours of sunlight, trophic status of the water body and pH can all influence the rate of degradation of the different herbicides. In terms of herbicidal effectiveness, degradation pathways are particularly important for products like fluridone or penoxsulam that require long exposure periods of 30 to 100 days. In these situations, the entire water body is often treated and therefore dispersion or dilution is not a problem, but the rate of degradation will often dictate product effectiveness. It is also important to mention the phenomenon of herbicide binding in relation to herbicide effectiveness. Several herbicides can bind with various ions in the water column, which can result in a reduction or loss of herbicidal activity. Binding is not a degradation pathway, but it can have an important influence on herbicide effectiveness. The best examples of product binding are the immediate binding of glyphosate to positively charged cations in the water column and the binding of diquat to negatively charged particles such as clay or organic matter in the water column. In both of these cases, the herbicide molecule remains intact but no longer has any herbicidal activity. The bound particles eventually settle to the sediments where microbial degradation takes place. Herbicides that are chemically bound in the sediment no longer have herbicidal activity and undergo microbial degradation over time.

The table on page 77 provides general information about exposure time requirements, typical aqueous half-lives that result from product degradation and the key degradation pathways for the registered aquatic herbicides.

Herbicide Concentration Monitoring

The above discussion of herbicide dissipation and half-lives is quite relevant to current use patterns of many aquatic herbicides. Operational monitoring of herbicide concentrations has increased significantly over the past 10 years. The advent of enzyme-linked immunoassays (ELISA) for several of the registered aquatic herbicides (including fluridone, endothall, triclopyr, 2,4–D and penoxsulam) has largely been responsible for this trend. While monitoring used to be very costly and was associated almost exclusively with regulatory studies or field research trials, several groups now offer monitoring support for operational treatments. When managers select herbicides such as fluridone and penoxsulam, the extended exposure requirements and large-scale use patterns are often supported by monitoring programs. In this case the monitoring can be used to manage the concentrations and exposure periods and to determine when and if additional herbicide applications are necessary to achieve optimal target plant control. In addition, monitoring can be used to determine when herbicide concentrations become low enough that use restrictions on water (e.g. irrigation, potable water use) can be lifted. There are numerous potential uses for operational monitoring of aquatic herbicide concentrations and given the value of the information that can be obtained, it is likely this trend will increase in the future.

Summary

This chapter lists eleven products that are registered by the EPA for aquatic plant control in aquatic systems. These herbicides are very different from one another; some have been used for decades, whereas others have only recently been approved for use in water. More specific directions regarding the use of these products are on the label and are also available from the companies that make, sell or distribute these herbicides.

For more information:
- Aquatic Plant Management Society. http://www.apms.org
- University of Florida Center for Aquatic and Invasive Plants. http://plants.ifas.ufl.edu
- US Army Corps of Engineers. http://www.erdc.usace.army.mil/pls/erdcpub/docs

Compound	General exposure requirements	Typical degradation half-life in water	Key degradation pathway and comments
Copper	Hours to 1 day	Hours to 1+ day	Copper is a natural element and is therefore not subject to degradation. Following application, copper ions are typically bound to particles or chemical ions in the water or sediment, which results in the loss of biological activity. Active copper ions in the water column are more readily inactivated in hard water systems. Concerns have been expressed regarding buildup of copper residues in sediments.
Endothall	Hours to days	2 to 14+ days	Endothall is a simple acid that is degraded via microbial action. Water temperature and the level of microbial activity can have a strong influence on the rate of degradation. Cooler water temperatures typically result in slower rates of degradation.
Diquat	Hours to days	½ to 7 days	Diquat is rapidly bound to negatively charged particles in the water column. Higher turbidity water can result in very fast deactivation of the diquat molecule. The ionic bonds between diquat and charged particles negate herbicidal activity. Once biologically inactivated, diquat is then slowly degraded via microbial action.
Carfentrazone	Hours to 1 day	Hours to 1 + day	Carfentrazone is degraded via hydrolysis. The rate of hydrolysis is pH–dependent, with faster degradation occurring in higher pH waters.
2,4–D	Hours to days	4 to 21+ days	The key degradation pathway for 2,4–D is via microbial degradation. Water temperature and rate of microbial activity can have a strong influence on the rate of degradation. Photolysis also plays a role in degradation.
Triclopyr	Hours to days	4 to 14+ days	The key degradation pathway for triclopyr is via photolysis or sunlight. Time of year, water depth and water clarity can influence the rate of photodegradation. There is also some microbial action that results in degradation.
Glyphosate	Not used for submersed	Hours to 1+ day	Glyphosate is rapidly deactivated once it contacts the water column due to immediate binding with positively charged ions in the water. Once bound to cations, glyphosate is biologically inactive. Microbial action ultimately degrades the glyphosate molecule in the sediment.
Imazapyr	Not used for submersed	7 to 14+ days	The key aqueous degradation pathway for imazapyr is via photolysis. Time of year, water depth and water clarity can influence the rate of photodegradation. Microbial degradation can also play a role.
Fluridone	45+ days	7 to 30+ days	The key degradation pathway for fluridone is via photolysis. Factors such as water depth, water clarity and season of application can influence photolytic degradation. Microbial activity can also play a supporting role in degradation.
Penoxsulam	45+ days	7 to 30+ days	The key degradation pathway for penoxsulam is via photolysis. Factors such as water depth, water clarity and season of application can influence photolytic degradation. Microbial activity can also play a supporting role in degradation.
Imazamox	14+ days	7 to 14+ days	The key degradation pathway for imazamox is via photolysis. Factors such as water depth, water clarity and season of application can influence photolytic degradation. Microbial activity can also play a supporting role in degradation.

Chapter 11

Chapter 12

Carole A. Lembi
Purdue University, West Lafayette IN
lembi@purdue.edu

The Biology and Management of Algae

Introduction

Algae are found in all salt and freshwaters worldwide. Although algae are very simple in their structure and sometimes consist only of a single cell floating in water, they are tremendously important for the health of our planet. Algae provide the base of food chains that support whales, seals, sharks and all other marine organisms in the oceans. In freshwaters, they also support food chains that lead to animals as diverse as bass, bald eagles and grizzly bears. Another essential role of algae is that they produce between 40-50% of the oxygen that we breathe through the process of photosynthesis!

The number of algae species is unknown, but it is likely more than 100,000, ranging from single cells to the large seaweeds found along our coastlines. Identification of freshwater algae can be difficult because the cells, or even clusters of cells, tend to be small and a microscope is usually required for accurate identification. In addition to cell shape and size, a key feature for proper identification is the color. Although all algae contain the green pigment chlorophyll, other pigments can also be present and can give the organisms different colors. Green algae are green because of chlorophyll, but diatoms and dinoflagellates are brown because xanthophyll pigments are present in higher concentrations than chlorophyll. The blue-green algae (also called the cyanobacteria) contain phycocyanin, a blue pigment that, along with chlorophyll, gives the cells a bluish-green color under the microscope.

Algae grow rapidly and reproduce primarily by cell division and by the formation of spores. They do not produce flowers or seeds. Most of the time people don't notice them, even though they are present in most bodies of water from bird baths to large lakes. Under certain circumstances, algae grow so prolifically that we do notice them. This is when water turns pea-soup green, or when masses of what is commonly called "moss" float on the surface of the water. It is these algae that often need to be managed because of the problems they can cause.

In addition to being unsightly, excessive algal growth (often called blooms) can lead to fish kills. This happens when the algae in a body of water die (crash) all at once.

Crashes can be caused by a variety of factors including cell aging, nutrient depletion or sudden changes in weather, such as a shift in water temperature or a period of prolonged cloudiness. Bacteria and fungi that break down the dead algal cells (organic matter) require large amounts of oxygen; as algae decompose, oxygen in the water is depleted, which results in oxygen-starved and dying fish. Since it is difficult to predict when and under what circumstances an algal bloom will crash, it is essential that waters be managed so that excessive growth does not occur. Once a body of water becomes infested with algae, control measures can be used to reduce the frequency and severity of blooms, but it is extremely difficult to eliminate the problem. Because of the many different types of algae and the need to initiate control measures so that fish kills do not occur, it is usually best to consult with a professional lake or pond manager for advice on management strategies. Hiring a certified aquatic pesticide applicator knowledgeable about algae control is also a good move when chemical treatments are recommended.

The algae that cause obvious changes to the color of the water itself are called phytoplankton. These algae consist of single cells or clusters of cells that can only be identified with a microscope. Another major group of algae forms long filaments, or strings, which get tangled together and form clumps or mats. Although these mats start growing along the bottom of a body of water, the oxygen they produce from photosynthesis gets trapped as air bubbles in the mats, causing them to detach from the bottom of the pond or lake and float to the surface. This is when mat-forming algae become visible and cause problems when people try to fish, swim or boat through the mats.

A third group of freshwater algae is the Chara-Nitella group. These algae look like flowering plants because they appear rooted and have "leaves" that are arranged along a stem. Chara, also called stonewort, usually grows in very hard water and is often calcified (covered with scale) and brittle, whereas Nitella tends to grow in softer waters. These algae provide valuable habitat for fish and stabilize sediments; however, in shallow water some species can grow to the surface and be troublesome.

Algae are usually identified by the taxonomic group to which they belong. From a management standpoint, the two major groups are blue-green algae and green algae. Phytoplanktonic blue-green algae are usually responsible for the pea-soup green color of water. These algae can be extremely harmful not only because they have the potential to cause fish kills by depleting oxygen when they die, but also because some produce toxic compounds that can poison livestock, pets and wild animals that drink contaminated water. In a few instances, humans have been sickened by drinking contaminated water; also, deaths have been recorded outside the United States. Such poisoning is very rare, but it is always wise to prohibit people from drinking or swimming in water that is dark green in color. Blue-green algae can also cause water to taste or smell foul and can cause fish flesh to taste musty.

Some filamentous blue-green algae form mats but most species of mat-formers are green algae. Mats that float on the surface often get "sunburned" from exposure to high light. The tops of the mats will look yellow; however, if the mat is pulled apart, the green color of the filaments or strings below the surface will be obvious.

Almost all of the algae that cause problems are native to the US and humans have been living with them for centuries. The conditions that promote algae include those typical of small, shallow ponds or lakes that become very warm in the summer and have little or no wave action. The main reason algae are such problems now is because of the impacts we humans have had on our water resources. Like other plants, algae require light, water and carbon dioxide to survive and grow. Light is seldom a problem in shallow waters. Algae and plants also need nitrogen, phosphorus and other nutrients in order to grow. The increase of nitrogen and phosphorus in lakes, rivers and ponds from many sources – including sewage and runoff from fertilized lawns, farm fields and livestock pastures – has caused algal blooms to proliferate in many bodies of water. Excessive algae growth is a key indicator of eutrophic conditions in lakes and ponds (see Chapter 1 for a discussion of trophic states). Even the Gulf of Mexico, which receives nutrient-laden waters from the Mississippi River, has suffered from algae blooms and fish kills.

What can be done to reduce the incidence and severity of an overabundance of algae?
Nutrient reduction and inactivation

A difficult but essential first step is to reduce the factors that cause algae to grow. This is most easily accomplished when constructing a body of water such as a pond. New ponds should be situated away from obvious sources of nutrients and dug deeply enough to prevent light from reaching the bottom. Unfortunately, reducing the input of nutrients into an established pond or lake can be quite difficult. Good watershed management plans are required to reduce obvious sources of nutrients such as upstream inputs from sewage outfalls, lawn or farm field fertilization and livestock operations. Every lake association should initiate and follow through on a watershed management plan. Nutrient sources from around the shoreline – including fertilization of lawns close to the water's edge – should be reduced as well. Fertilization should be prohibited within at least 10 to 20 feet of the shoreline and fertilizers without phosphorus should be used in areas that have to be fertilized.

Turfgrasses are usually maintained along the shoreline, but these grasses have shallow roots and do little to prevent erosion. Recent interest has focused on planting shorelines with native emergent vegetation such as sedges, rushes and colorful plants such as pickerelweed, cardinal flower and arrowhead. These native plants, which are sold by companies promoting environmental restoration, have longer and more substantial root systems than turfgrasses, which allows them to hold soil better, prevent erosion and potentially absorb more nutrients from subsurface runoff.

Some nutrient inactivation methods in the water itself can help reduce algae blooms. Alum is a material that combines with phosphorus and causes it to precipitate to the bottom so that it is no longer available for algal growth. However, the long-term value of an alum application can be greatly reduced if inputs of phosphorus from the shoreline and watershed continue unabated. Also, alum lowers the pH of the water, which can be detrimental to fish life. Buffers are usually added to prevent this, so the application of alum is best left to an experienced contractor.

Another option is to install aerators. The introduction of oxygen into a body of water changes the chemistry of the water so that phosphorus is precipitated to the bottom. Aeration is also valuable for fish life and the introduction of air (oxygen) to the water promotes the bacterial breakdown of organic matter that has accumulated on the bottom over time. Fountains that just spray water from the pond surface are not effective aerators because they only aerate the top few feet of water. Effective aeration devices are those that deliver oxygen to the bottom waters. They can be purchased or constructed and work on one of two principles. One is to pump air into weighted tubes along the pond bottom. The oxygen bubbles into the water through holes in the tubes. This is the most commonly used device in ponds. The second type of aerator, called a hypolimnetic aerator, moves low-oxygen bottom water to the surface, oxygenates it and then recirculates the aerated water to the bottom of the lake. These units are typically used on stratified lakes where the bottom waters are cold and the aerated cold water must be returned to the bottom in order to support cold-water fish.

Several enzyme and bacterial products are on the market and claim to reduce the amount of nutrients available to algae. The enzymes are thought to break down organic matter so that it is easier for the natural bacteria to take up the nitrogen and phosphorus that is released during decomposition of the organic matter. Adding a product that contains bacteria is intended to supplement the natural bacteria population. In theory, bacteria are better competitors for nutrients than are algae. Consequently, the bacteria should reduce the amount of nitrogen and phosphorus that is available for algae growth, resulting in clear water. Unfortunately, very little research has been conducted on the effectiveness of these products and testimonials are mixed, so their usefulness is controversial.

Nutrient reduction/inactivation strategies can help improve the overall health of a body of water. On the other hand, they seldom cure algae problems because it is usually difficult to identify the source of inputs of nitrogen and phosphorus. Is it lawn fertilization? Is it from the recycling of nutrients from the lake or pond sediments? Is the soil naturally rich in nutrients? Or is it from a number of other potential sources? Without this information, it is difficult to develop a nutrient reduction strategy that results in relatively rapid and long-term control of algae.

Other control options

Reducing light penetration through the use of EPA-registered dyes can be helpful in algae control. Dyes should be applied early in the growing season before algae appear at the surface. However, since algae often start growing in shallow water, the dye may not be at a high enough concentration in those areas to sufficiently reduce algal populations. Once algae begin to grow in shallow water, they can then spread to the upper portions of the deeper water relatively quickly. Since the dye concentration in the water must be maintained throughout the growing season, dyes are more effective on bodies of water that have little to no outflow. Dyes alone are seldom effective for controlling algae, but they can be used after an algicide treatment to reduce regrowth.

Mat-forming algae can be raked out manually or with mechanical harvesters. Raking is typically done around boat docks and in swimming areas. Since mat-formers are mostly free-floating, new mats can rapidly reinfest an area that has been raked.

The only biological control agent (Chapter 8) being used for algae control is the tilapia (*Tilapia zillii*), a fish that has been introduced into and can only survive in waters of the southern US. Tilapia are stocked in very high numbers in the cooling reservoirs of some southern power plants, but they are not used by the public. The grass carp or white amur (Chapter 10) does not feed on phytoplankton. When young, grass carp will consume some mat-forming algae, but they do have preferences (slimy algae are rejected; coarser algae might be eaten). As the grass carp age, they tend to feed more on submersed plants than on algae.

Algicides

Direct control of algae is most frequently accomplished with algicides. Copper sulfate has been used for algae control since the early 1900s and is used on more surface acres of water than any other product that controls algae or aquatic plants. One of the benefits of copper sulfate is that phytoplanktonic blue-green algae are more sensitive to it than are phytoplanktonic green algae. As a result, noxious blue-green algae can often be removed without harming the green algae, which are usually desirable because they are an important component of the aquatic food chain. Both copper sulfate and the copper chelated products are also used to control mat-forming algae. Liquid formulations of chelated copper products are particularly effective for this purpose because they can be easily mixed with water and sprayed directly onto the algae mats.

Copper sulfate and copper chelates are widely used throughout the world to treat reservoirs that collect and store drinking water. Our ability to safely treat water with copper products to control blue-green and other algae is predicated on the low dosages used, the fact that copper precipitates out of the water and into the sediments within several days in moderately hard to hard waters, and on the inability of copper to bioaccumulate (build up over time) in fauna in the food chain. Animals and humans actually require small amounts of copper in their diets and the element is often included in human vitamin supplements and in animal feed. Copper from treated water that is consumed by humans and other animals passes through the body and is expelled in the urine rather than moving into the body's tissues. Copper products can be applied to water with no restrictions on water use (e.g., swimming, fishing, drinking); however, they should be used very carefully or not at all in waters that contain sensitive fish species such as trout, koi and goldfish.

Copper products are effective and widely used, but they do not solve the underlying issue of why the algae are there in the first place. These algicides do offer short-term relief, which can be extremely valuable in terms of preventing fish kills (if treatment is initiated before the bloom becomes severe) and opening up the water for fishing, swimming and other activities. However, it is extremely unlikely that copper applications will kill all the algae or their spores, so regrowth almost always occurs. Furthermore, copper products are very short-lived in the water and algae can start to reappear quickly, sometimes within several weeks. As a result, the potential for retreatment has to be part of any management plan that uses copper products.

There are very few alternatives to copper for direct algae control. The amine salt of endothall has algicidal activity and can be sprayed along the edges of ponds for control of mat-forming algae. Read and follow the herbicide label for endothall carefully as this herbicide can be toxic to fish if not used correctly. Compounds that are based on sodium carbonate peroxyhydrate release hydrogen peroxide into the water, which rapidly kills the algal tissue it comes into contact with. Unlike copper, hydrogen peroxide breaks down rapidly in water to produce hydrogen and oxygen, so it leaves no residues. A uniform application of the sodium carbonate peroxyhydrate granules is necessary to ensure optimum results because the hydrogen peroxide products only control algae that come into direct contact with the granules. Since hydrogen peroxide products are fairly new to the market and have not been available for very long, they have not been tested for effectiveness as extensively as copper. Research is still needed to determine which algal species are most effectively controlled by these products.

Another chemical approach that has received much publicity is the use of barley straw for algae control. English researchers found that bundles of barley straw placed in water released a toxin that killed algae as the straw decomposed. A number of barley products, including barley straw extracts, are on the market. The potential of the toxin to kill algae is well established but the conditions under which the activity occurs are as yet unknown. In other words, we do not know which algal species are affected nor do we know what effects water temperature, water hardness, nutrient status, etc. might have on the effectiveness of this treatment. Anecdotal evidence suggests that the method is inconsistent; that is, it might work on one body of water but not on another, and the reason for this is not known. Caution, along with much reading and study, are recommended before attempting to use barley straw to control algae.

Summary

Algae problems are usually the result of too many natural- or human-derived nutrients in a body of water. As long as light, nutrients and water are available, something green will grow. Even swimming pools can develop algae problems because different types of algae have different nutrient requirements and all water – even rainwater – contains nutrients. The algae that cause most problems are blue-green algae and mat-forming green algae. Due to their diversity and ability to reproduce quickly, algae are difficult to control. Many products claim to reduce algal populations, but unless they make direct claims of algae control, they do not have to be registered for use with the EPA and are largely untested. Products that are registered with the EPA include some dyes and algicides such as the copper, peroxide and endothall products. Specific use directions are explicitly stated on the labels, which are excellent sources for further information.

For more information:

- Algae control with barley straw (Ohio State University Extension Fact Sheet)
http://ohioline.osu.edu/a-fact/0012.html
- Algae: some common freshwater types (Microscopy UK)
http://www.microscopy-uk.org.uk/index.html?http://www.microscopy-uk.org.uk/pond/algae.html
- Blue-green algae (cyanobacteria) blooms (California Department of Public Health)
http://ww2.cdph.ca.gov/healthinfo/environhealth/water/Pages/bluegreenalgae.aspx
- Blue-green algae photo gallery (Vermont Department of Health)
http://healthvermont.gov/enviro/bg_algae/photos.aspx
- Harmful algal blooms (HABs) (Centers for Disease Control and Prevention)
http://www.cdc.gov/hab
- Identifying and managing aquatic vegetation (Purdue University)
http://www.extension.purdue.edu/extmedia/APM/APM_3_W.pdf
- Plant identification: algae, AQUAPLANT (Texas A&M University)
http://aquaplant.tamu.edu/database/index/plant_id_algae.htm
- Surf your watershed (United States Environmental Protection Agency)
http://cfpub.epa.gov/surf/locate/index.cfm

Photo and illustration credits:

Page 79: Algae bloom; Carole Lembi, Purdue University
Page 80: Chara; Lyn Gettys, University of Florida Center for Aquatic and Invasive Plants
Page 82 Aerator; William Haller, University of Florida Center for Aquatic and Invasive Plants
Page 83: Filamentous algae; Andy Price
Page 84: Algae bloom; Carole Lembi, Purdue University

Chapter 12

Chapter 13

William T. Haller
University of Florida, Gainesville FL
whaller@ufl.edu

Introduction to the Plant Monographs

The ten aquatic and wetland plants described in this chapter have one thing in common: all are of foreign origin. In addition, most were intentionally introduced to North America by humans. While native aquatic plants can sometimes become problematic, the plants in this chapter have caused significant economic and ecological damage to ecosystems throughout North America and will continue to do so for the foreseeable future. If you live in an area where none of these plants are found, you are among the fortunate few. You and your neighbors should make every effort to prevent the introduction and movement of these noxious weeds in your area.

The authors of the following plant descriptions have devoted years to researching the biology and control of these invasive species. Each weed species included in this chapter has distinct characteristics that cause it to be invasive and requires different techniques for control, but all authors agree on one concept – prevention is the most efficient and cost-effective method to protect natural areas from invasion by these noxious species.

A wealth of information is available on the internet about invasive species in general and the species described in this chapter. Excellent reference sources include local sites such as your state environmental protection agency and state invasive species working groups. National resources include the following websites:

The United States Environmental Protection Agency
http://www.epa.gov/

The University of Florida Aquatic Plant Information Retrieval System Online Database
http://plants.ifas.ufl.edu/APIRS

The University of Florida Center for Aquatic and Invasive Plants
http://plants.ifas.ufl.edu

USDA NRCS. The PLANTS Database. National Plant Data Center, Baton Rouge, LA
http://www.plants.usda.gov/

US Army Corps of Engineers Aquatic Plant Information System
http://el.erdc.usace.army.mil/aqua/apis/apishelp.htm

Information and knowledge are the keys to prevention. Familiarize yourself with the characteristics of the invasive species described in this chapter so that you can positively identify them in the field. If you encounter a new population of one of these weeds, immediately notify the appropriate agency in your state and provide them with as much information as possible, including the location of the population. We are all responsible for the protection and stewardship of the ecosystem and your attention to detail can play a critical role in preventing the spread of these invasive species.

Chapter 13

William T. Haller
University of Florida, Gainesville FL
whaller@ufl.edu

Chapter 13.1: Hydrilla

Hydrilla verticillata (L.f.) Royle; submersed plant in the Hydrocharitaceae (frog's-bit) family
Derived from *hydr* (Greek: water) and *verticillus* (Latin: whorl)
"water plant with whorls of leaves"

Introduced from Asia to Florida in the late 1950s
Present throughout the southeast and north to New England and Wisconsin; west to California, Washington and Idaho

Introduction and spread

Hydrilla (*Hydrilla verticillata*) is the only species in the genus *Hydrilla* but several biotypes occur in its native range. Some biotypes are monoecious (each plant has both male and female flowers) and others are dioecious (each plant bears only male or female flowers). It appears that hydrilla was introduced to North America on at least two separate occasions, which accounts for the distribution of two biotypes in the United States. The monoecious biotype was introduced most recently and may be the more cold-tolerant of the types. It was first discovered in the Potomac River in the late 1970s and can now be found in most areas north of Lake Gaston on the NC/VA border. The female-flowering dioecious biotype was introduced earlier and occurs exclusively in the southern United States from Florida to North Carolina and west to Texas. Dioecious hydrilla was introduced into Florida by the aquarium nursery trade in the 1950s and was spread rapidly throughout the state by intentional (plant growers) and unintentional (boat trailers) means. By the late 1970s hydrilla was included on the Federal Noxious Weeds List and a number of state prohibited plant lists as well. These listings have stopped the interstate sale and shipping of the species, but hydrilla is continually spread by irresponsible boaters and others who move plants from one watershed to another, since the species easily reproduces and forms new colonies from small plant fragments. There has been no direct evidence to suggest that hydrilla is spread by waterfowl and other aquatic fauna, but this type of transfer may occur between bodies of water that are in close proximity to one another. The introduction and spread of monoecious hydrilla in the northern US has not been well-documented because its appearance is very similar to that of the native elodea (*Elodea canadensis*). However, many confirmed initial infestations have occurred near public access points, suggesting that boaters continue to inadvertently transfer hydrilla on trailered boats.

Description of the species

Hydrilla is a rooted submersed perennial monocot that grows in all types of bodies of water, with its growth limited only by water depth and velocity of flow. The stems of hydrilla are slender (about 1/32" in thickness), multi-branched and up to 25 feet in length – stems can grow as much as an inch per day. Hydrilla forms dense underwater stands and often "tops out" to form dense canopies or mats on the surface of the water. All vegetative parts of hydrilla are submersed and the appearance of the species can vary drastically depending on growth conditions such as water pH, hardness and clarity.

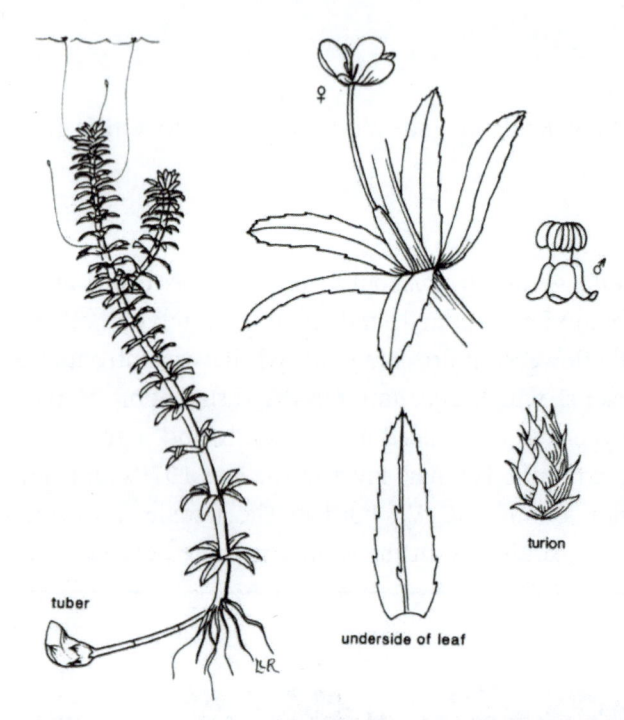

Hydrilla has small (to 5/8" in length), strap-like, pointed leaves. The midrib on the underside of the leaf often has one or more sharp teeth along its length and leaf margins are distinctly saw-toothed, especially in hard water. Leaves are attached directly to the stem and are borne in whorls of four to eight around the stem, with a space of 1/8" to 2" between whorls. Healthy leaves are bright green, whereas leaves under stress from fungi, bacteria and sun-bleaching may be brown or yellow. Hydrilla is often confused with native elodea and exotic egeria or Brazilian elodea (Chapter 13.8). While these three species are very similar in appearance, leaves of native elodea are borne in whorls of three and those of egeria are arranged in whorls of four or five. In addition, only hydrilla has saw-toothed leaf margins; the leaf margins of the other species are smooth. It is often difficult – even for trained biologists – to tell hydrilla, native elodea and egeria apart. Plants can be positively identified as hydrilla by digging 1 to 2" into the soil and looking for the presence of tubers or turions among the roots, as hydrilla is the only one of these species to produce these reproductive structures.

Reproduction

Dioecious hydrilla can only spread by vegetative means such as plant fragments because it does not produce seeds. Its spread by this method has been rapid and has increased the species' range throughout most of the southeastern US. Hydrilla produces two types of vegetative reproductive structures: turions and tubers. Turions are small (to 1/4" in diameter), cylindrical, dark green and borne in leaf axils, whereas tubers are larger (to 1/2" in diameter), potato-like, yellowish and attached to the tips of underground rhizomes 1 to 3" below the surface of the sediment. Dioecious hydrilla produces tubers and turions during winter short-day conditions in the southeastern US, whereas monoecious hydrilla behaves like an annual and produces these structures in mid to late summer in northern waters. Hydrilla is the only species in the Hydrocharitaceae family to produce tubers and turions, so the presence of these structures is considered confirmation that the plant in question is indeed hydrilla. Underground tubers can remain dormant for many years; this protects the species from management efforts such as drawdowns and allows plants to survive adverse conditions. Studies have shown that a single sprouting tuber of dioecious hydrilla planted in shallow water can produce over 200 tubers per square foot each year.

The ecological importance of sexual reproduction in monoecious hydrilla (with both male and female flowers) is unknown. Flowers and seeds of hydrilla are tiny and therefore difficult to study in natural systems, but viable seeds have been produced under experimental conditions. Dioecious plants produce only female flowers and the lack of male flowers for pollination prevents seed formation. The female flowers of hydrilla are tiny (up to 1/16" in length), white and borne singly on threadlike stalks. These stalks are attached to the stem in leaf axils near the tip of the stem and are up to 4" in length, which allows the flowers to be level with the surface of the water. Male flowers are tiny, greenish and closely attached to leaf axils near the stem tips. When ripe, they separate from the stem and float to the surface, where they pollinate the female flowers by randomly bumping into them and dropping pollen into the female flower.

Problems associated with hydrilla

Hydrilla grows almost entirely underwater as a submersed aquatic plant and its growth potential is limited primarily by water clarity and depth of light penetration. Hydrilla has been reported at depths of 35 to 40 feet in crystal clear spring water and is commonly found at water depths of 15 to 20 feet in lakes with clear water. Hydrilla is uniquely adapted to grow under low light conditions, which allows it to colonize water that is deeper than most native submersed species can tolerate. For example, native submersed plants typically colonize the margins of shallow lakes where water depth is 6 to 8 feet. Hydrilla competes with native plants in these shallow areas, but also grows in much deeper water with no competition, which greatly extends the spread of the vegetated littoral zone outward from the shoreline.

Hydrilla infestations often go unnoticed until the species "tops out" and reaches the surface of the water, where it forms hundreds of lateral branches due to the increased light intensity. This surface canopy or mat formed in the upper 1 to 2 feet of water comprises as much as 80% of the biomass of the plant on an area basis and limits light availability to lower-growing native submersed plants, which reduces species diversity over time. The ecological effects of this dense growth on the water surface include significant changes in water temperature, wave action, oxygen production, pH and other parameters, which reduce the suitability of infested waterways for use by aquatic fauna. Human activities are adversely affected as well – recreational use of water is limited, property values are diminished and there are increased public health and safety concerns (e.g., mosquito control, drowning, flooding). The severity of problems caused by hydrilla depends on the characteristics of the infested water body. An acre or two of hydrilla in a 100-acre lake may cause few problems; however, coves, bays or lakes with infestations of 80% or greater are significantly impacted by hydrilla.

Management options

Clearly, preventing hydrilla from entering a water body is the best method to control this noxious species. Federal and state authorities have made it illegal to sell and transport hydrilla, which reduced this source of infestation. However, hydrilla still manages to increase its range and to colonize new bodies of water. Once hydrilla becomes established in a water body, control options are costly and generally must be employed on an annual basis.

Mechanical (Chapter 7) or physical (Chapter 6) control projects such as hand removal, benthic barriers or mechanical harvesters should be designed to prevent the spread of hydrilla fragments to other parts of the water body. Of course, if a lake is already extensively infested by hydrilla, there is less concern regarding plant fragmentation. Hand removal is labor-intensive and must take into consideration the presence of tubers and turions in and on the sediment, since failure to remove these structures virtually assures rapid reinfestation of the site. Mechanical harvesting can be expensive and most harvesters only cut to a water depth of 5 feet. Since hydrilla can grow an inch per day, control may only last for 2 months after mechanical harvesting. Another problem associated with mechanical harvesting is disposal of the harvested hydrilla. This vegetation has been evaluated for its potential as mulch, cattle feed, biofuel production and other uses, but its utility is very limited. Also, submersed plants do not produce much dry matter – a surface mat of hydrilla may weigh as much as 15 tons per acre, but contains only 5% (1,500 pounds) dry matter. As a result, harvested hydrilla is generally disposed of in a landfill due to its high water content (95% by weight) and low production of biomass.

Drawdowns and freezing of hydrilla tubers and turions may provide temporary control in northern locations, but these measures provide only a season or partial season of control in the southeastern US. Thus, most hydrilla management programs rely on the use of biological control agents (grass carp) or herbicides.

Classical insect-based biocontrol of hydrilla has been studied for at least 30 years. Researchers continue to seek possible biocontrol insects, pathogens and other agents in Asia and Africa. A few promising candidate insects have been discovered, studied and released to control hydrilla, but these insects have provided only localized and temporary reductions in hydrilla populations and are not considered to be viable biocontrol agents (Chapter 9).

In contrast, sterile triploid grass carp (Chapter 10) are widely used for hydrilla control in some states. Grass carp are released primarily in closed ponds or lakes and are sometimes used in conjunction with herbicides. Grass carp are not species-specific as required for the introduction of biocontrol insects; grass carp may prefer hydrilla but will consume most submersed and emergent aquatic plants. As a result, most states regulate the stocking and use of grass carp. Despite this challenge, grass carp continue to be the most effective method for biological control of hydrilla where their use is legal and practical.

Several herbicides can be used to effectively control hydrilla, but one of the most significant problems associated with chemical control of any submersed species is dilution (Chapter 11). An acre of water that is 1 foot deep comprises 325,800 gallons of water, which results in tremendous dilution of herbicides. In addition, water flow or movement greatly reduces the amount of time hydrilla is exposed to the herbicide. These factors can make it difficult to control hydrilla using chemical methods, so treatments should be designed to take dilution and water movement into consideration. Fast-acting contact herbicides – including copper, diquat and endothall formulations – are taken up quickly by hydrilla and result in rapid plant death and decay. These herbicides are generally used for spot treatments, strip treatments along shorelines and where water movement would limit use of slower-acting systemic herbicides.

Slow-acting systemic herbicides – including fluridone, imazamox and penoxsulam – control hydrilla by inhibiting enzyme activity. These herbicides are usually applied as whole-lake treatments and provide control of hydrilla only when a long period of contact is possible. An advantage to systemic herbicides is that they are effective at low rates – usually concentrations of less than 100 ppb or even less than 10 ppb in the cases of fluridone and penoxsulam. These herbicides slowly kill plants by starving them over a long period of time, but usually provide 1 to 2 years of control. Slow plant decay resulting from systemic herbicide treatments minimizes possible oxygen depletion and reduces the potential for fish mortality. The disadvantage of systemic herbicides is that they generally require a total lake treatment, or at least treatment in coves, bays and other areas where water movement and dilution are reduced and there is little or no water exchange. Most states require permits to apply herbicides in public (and some private) waters, so contact your state water authority for further advice and information.

Summary

Prior to 1950 there was no scientific information suggesting that hydrilla would cause such serious problems throughout the world. Hydrilla has become one of the world's worst submersed weeds as water resources have been developed and it now causes problems in all tropical and subtropical continents with the exception of Africa. Hydrilla has spread from Florida north to Maine and Wisconsin and northwest to Washington in the span of only 50 years. The annual cost to control hydrilla in public waters in Florida alone totals approximately $15 million. Florida is particularly impacted by hydrilla due to its moderate climate and shallow, naturally nutrient-rich lakes, but research on the distribution of hydrilla in Asia predicts that hydrilla could colonize virtually any area in North America and could survive as far north as Hudson Bay.

For more information:
•Langeland KA. 1996. *Hydrilla verticillata* (L.f.) Royle (Hydrocharitaceae): the perfect aquatic weed. Castanea 61:293-304. http://plants.ifas.ufl.edu/node/184
•Madeira PT, CC Jacono and TK Van. 2000. Monitoring hydrilla using two RAPD procedures and the nonindigenous aquatic species database. Journal of Aquatic Plant Management 38:33-40.
http://apms.org/japm/vol38/v38p33.pdf
•McLane WM. 1969. The aquatic plant business in relation to infestations of exotic aquatic plants in Florida waters. Journal of Aquatic Plant Management 8:48-49.
http://apms.org/japm/vol08a/v8p48.pdf

Photo and illustration credits:
Page 89: Hydrilla infestation; Vic Ramey, University of Florida Center for Aquatic and Invasive Plants
Page 90: Line drawing; University of Florida Center for Aquatic and Invasive Plants
Page 91: Hydrilla bouquet; William Haller, University of Florida Center for Aquatic and Invasive Plants

John D. Madsen
Mississippi State University, Mississippi State MS
jmadsen@gri.msstate.edu

Chapter 13.2: Eurasian Watermilfoil

Myriophyllum spicatum L.; submersed plant in the Haloragaceae (watermilfoil) family
Derived from *myrios* (Greek: numberless), *phyllon* (Greek: leaf) and *spica* (Greek: spike)
"plant with many leaf divisions that bears flowers in a spike"

Introduced to several locations in the US from Europe in the 1940s
Present throughout the continental US and Alaska

Introduction and spread

Eurasian watermilfoil (*Myriophyllum spicatum*) is one of fourteen species of *Myriophyllum* present in the US. Most species of this genus in the US are native, but two (*M. aquaticum* and *M. spicatum*) are exotic species that have been introduced to North America. Of these two exotic species, Eurasian watermilfoil is much more widespread and more problematic. The species was first reported in the US in the 1940s and spread rapidly into the mid-Atlantic and midwestern states in the 1960s and 1970s. Eurasian watermilfoil also became a serious problem in the hydropower and flood control reservoirs of the Tennessee River, where large-scale applications of herbicides were used in an attempt to eradicate the weed. Eurasian watermilfoil is still present in the TVA (Tennessee Valley Authority) system but has largely been displaced by hydrilla (Chapter 13.1). More recently (from the 1980s until 2009) the species has invaded lakes in Idaho, Minnesota and Maine and continues to expand its coverage throughout the northern US. Eurasian watermilfoil is now the most widespread submersed aquatic weed in the northern half of the US.

Eurasian watermilfoil has been introduced to the US multiple times and was likely first brought to North America in ship ballasts or as an ornamental plant for aquariums or water gardens. Accidental spread of Eurasian watermilfoil within the US is due primarily to transportation of contaminated boat trailers, boat parts and bait containers, but the species is also spread through the aquarium trade. Once Eurasian watermilfoil is introduced to a water system, it spreads prolifically by stem fragments that are produced both naturally (when stem sections detach from the plant at abscission sites) and as a result of mechanical breakage (when plants come into contact with boat motors and intense wave action). Some researchers speculate that Eurasian watermilfoil may be spread by wildlife or waterfowl; however, no direct evidence exists to support this theory. Eurasian watermilfoil produces numerous viable seeds, but the seeds contribute little to the propagation and spread of the plant. Eurasian watermilfoil was too widespread to be listed as a Federal Noxious Weed when the list was first developed; however, the species is listed on numerous state noxious and prohibited plant lists.

Description of the species

Eurasian watermilfoil is rooted in the sediment and grows completely underwater as a submersed plant that forms a dense canopy on the water surface. The species is commonly found in water from 1 to 15 feet in depth but can occur at depths of up to 30 feet if the water is extremely clear. Eurasian watermilfoil is an evergreen perennial plant that produces persistent green shoots throughout the year and overwinters as root crowns. Leaves are pinnately compound (feather-like), with each leaf composed of 14 to 24 pairs of leaflets arranged in whorls (groups) of four at the nodes of the stem. Stems and plant tips may appear reddish, but color is not consistent and may vary based on a number of factors, including environmental conditions. Flowers form on short aerial stems that hold them above the water and have both pollen-bearing ("male") and seed-producing ("female") flowers. Flowers are wind-pollinated and produce up to four nutlets per flower. Eurasian watermilfoil is difficult to identify and is often confused with several native species of *Myriophyllum*, including northern watermilfoil (*M. sibiricum*) and whorled watermilfoil (*M. verticillatum*). Hybridization between Eurasian and northern watermilfoils reportedly occurs in the field and the seedlings produced from these cross-pollinations often have features that are intermediate to the parental plants.

Reproduction

Eurasian watermilfoil produces a significant number of viable seeds and plants can be propagated from seed in the laboratory or greenhouse. However, successful colonization of new plants from seed in nature has not been documented. As a result, sexual propagation is generally thought to play an insignificant role in the spread of Eurasian watermilfoil. The species reproduces predominantly by vegetative means through fragmentation, which occurs when stems are broken mechanically (from wave action or contact with boat motors) and when stem sections naturally abscise or detach from the plant. Stem sections that result from natural breakage have high concentrations of starch and are likely responsible for most of the spread of the species. Eurasian watermilfoil can also spread by forming new root crowns on runners, which are produced when stems arch down, come into contact with the sediment and form roots that create a new root crown. Root crowns can also spread through the formation of rhizomes under the sediment, although detailed studies of this process have not been conducted. Root crowns overwinter and produce new shoots every year. As a result, more stems are added to root crowns each year, which increases stem density in the water column.

Problems associated with Eurasian watermilfoil

Because Eurasian watermilfoil grows entirely underwater as a submersed aquatic plant, the range of water depths the species can inhabit is limited by light penetration and water clarity. A dense canopy often forms at the surface of the water, which interferes with recreational uses of water such as boating, fishing and swimming. Dense growth of Eurasian watermilfoil may also obstruct commercial navigation, exacerbate flooding or clog hydropower turbines. In addition, excessive growth of the species may alter aquatic ecosystems by decreasing native plant and animal diversity and abundance and by affecting the predator/prey relationships of fish among littoral plants. A healthy lake is damaged because heavy infestations of Eurasian watermilfoil lower dissolved oxygen under the canopy, increase daily pH shifts, reduce water movement and wave action, increase sedimentation rates and reduce turbidity.

Management options

Prevention is always the best option to avoid infestations of Eurasian watermilfoil. Posting signs at boat launches and requesting that lake users watch for Eurasian watermilfoil and remove all plant material from boats before launching can be a successful strategy. When prevention methods are unsuccessful, early detection and rapid response to new infestations have been shown to reduce management costs over the long term.

There are currently no biological control agents that effectively control Eurasian watermilfoil. For example, grass carp (Chapter 10) do not feed on this species. Numerous studies have been conducted to evaluate the utility of native insect herbivores as potential biocontrol agents of Eurasian watermilfoil, but none have proven to be predictable and effective to date. Also, if native insects were able to effectively control introduced populations of Eurasian watermilfoil, new introductions of the weed would not result in population development and expansion to weedy proportions. Historical accounts of the introduction and spread of Eurasian watermilfoil suggest this is has not occurred. In addition, the use of native insects as biocontrol agents remains controversial (Chapter 8).

Several herbicides can be used to effectively manage Eurasian watermilfoil. Contact herbicides – including diquat and endothall – provide good control, whereas systemic herbicides such as 2,4–D, fluridone and triclopyr provide excellent control. Herbicides should be selected based on site size and conditions, water exchange characteristics, potential water use restrictions, federal, state and local regulations and economic considerations (Chapter 11).

Mechanical controls (Chapter 7) are also widely used to control small infestations of Eurasian watermilfoil. Mechanical harvesting and raking provide temporary but fair control in bodies of water that are small to moderate in size, whereas hand harvesting and suction harvesting provide longer term control than mechanical harvesting or raking. None of these mechanical methods alone results in long-term control of Eurasian watermilfoil; as such, these methods should be employed as part of an integrated weed control strategy.

Physical control techniques such as drawdowns, dredging and bottom barriers (Chapter 6) can reduce or prevent growth of Eurasian watermilfoil by altering the environment. Drawdowns require dewatering of the affected lake or pond and are particularly effective during the winter. Draining the water out of the system exposes the root crowns of Eurasian watermilfoil to the air and results in desiccation and death of the plants. Dredging is expensive but results in water depths too great for plants to grow. Dredging provides multi-season control but should only be used as part of a broader lake restoration effort. Bottom barriers are semi-impermeable sheets of synthetic material that are placed over the plant bed, which kills the plants underneath. Bottom barriers are expensive but can provide effective control of Eurasian watermilfoil in small areas.

Summary

Eurasian watermilfoil is an exotic aquatic weed that is widely distributed throughout North America. The species is most commonly associated with problems in temperate lakes, but invades tidal estuaries, saline prairie lakes, rivers and southern reservoirs as well. Although the economic impact of Eurasian watermilfoil is not as great as that of hydrilla or waterhyacinth (Chapter 13.5), its geographic and ecological distribution surpasses that of other North American aquatic weeds. In fact, problems associated with Eurasian watermilfoil are significant enough that states such as Idaho, Minnesota, Vermont and Washington have developed specific management programs to control invasions of Eurasian watermilfoil.

For more information:

•Grace JB and RG Wetzel. 1978. The production of Eurasian watermilfoil (*Myriophyllum spicatum* L.): A review. Journal of Aquatic Plant Management 16:1-11.
http://www.apms.org/japm/vol16/v16p1.pdf

•Jacono CC and MM Richerson. 2003. *Myriophyllum spicatum* L. Nonindigenous Aquatic Species web page, U.S. Geological Survey, Gainesville FL. http://nas.er.usgs.gov/plants/docs/my_spica.html

•Madsen JD. 2005. Eurasian watermilfoil invasions and management across the United States. Currents: The Journal of Marine Education. 21(2):21-26.

•Smith CS and JW Barko. 1990. The ecology of Eurasian watermilfoil. Journal of Aquatic Plant Management 28:55-64. http://www.apms.org/japm/vol28/v28p55.pdf

Photo and illustration credits:

Page 95: Eurasian watermilfoil infestation; Ryan Wersal, Mississippi State University Geosystems Research Institute
Page 96: Line drawing; University of Florida Center for Aquatic and Invasive Plants
Page 97: Eurasian watermilfoil; John Madsen, Mississippi State University Geosystems Research Institute
Page 98: Eurasian watermilfoil; John Madsen, Mississippi State University Geosystems Research Institute

Scott A. Kishbaugh
New York State Department of Environmental Conservation, Albany NY
sakishba@gw.dec.state.ny.us

Chapter 13.3: Waterchestnut

Trapa natans L; floating-leaved plant in the Trapaceae (waterchestnut) family; originally placed in the Hydrocharitaceae (frog's-bit) family; sometimes placed in the Lythraceae (purple loosestrife) family
Derived from *calcitrapa* [Latin: a spiked iron ball ("caltrops") used as an ancient weapon] and *natans* (Latin: swimming)

Introduced from Asia to Massachusetts and New York in the late 1870s to early 1880s
Present in the mid-Atlantic into the Northeast, south to northern Virginia, west to central Pennsylvania, east to New Hampshire, north to Quebec

Introduction and spread

Some botanists have subdivided the genus *Trapa* into more than 25 different species based upon small differences in the nutlets. Under the most recent taxonomic schemes, *Trapa natans* is subdivided into three varieties. The varieties *Trapa natans* var. *bispinosa* and *Trapa natans* var. *bicornis* are found primarily in northern India and southeastern Asia, where both are grown as agricultural crops, whereas the variety *Trapa natans* var. *natans*, commonly called waterchestnut, is a prized agricultural crop in India and China, a protected and disappearing plant in Europe and a highly aggressive invader in the United States. Waterchestnut is often confused with the Chinese waterchestnut (*Eleocharis dulcis*), an edible tuber common in Chinese cuisine. Both species have been widely cultivated as a food source, but they are unrelated. Although "waterchestnut" is the most widely used common name for *Trapa natans* var. *natans*, the variety is also known by a number of other common names, with religious ("Jesuit's nut"), evocative ("water caltrops") and sinister ("devils nut", "death flower") connotations.

Waterchestnut is native to Eurasia and Africa and archaeologists have found evidence of waterchestnut in sediments dating back to at least 2800 BC. The first introduction of waterchestnut to the US is better documented than that of most other exotic plants, but there is some debate regarding the specific time and place of this introduction. The initial introduction to North America was well-described by Eric Kiviat in a Hudsonia newsletter. North American infestations can probably be traced to two distinct locations. Waterchestnut was first introduced from Europe to Middlesex County, Massachusetts around 1874 and was cultivated as an ornamental in Asa Gray's botanical garden at

Harvard University in 1877. Seeds were distributed by Harvard gardeners into nearby ponds over the next several years; as a result, waterchestnut migrated into the Concord and Sudbury Rivers by the mid 1880s, reached nuisance portions by the turn of the century and underwent explosive growth by the 1940s.

Another introduction occurred in Scotia in eastern New York during the early 1880s. A Catholic priest planted waterchestnut seeds from Europe in Sanders Pond (now Collins Lake), which led to extensive colonization of the lake by 1884. Subsequent flooding of the neighboring Mohawk River (via locks and dams on the New York Barge Canal) further spread the plant and spawned widespread growth by the 1920s. Waterchestnut was reported in the Hudson River by 1930 and reached nuisance levels in the 1950s. The species likely then spread west through the Erie Barge Canal system and reached Oneida Lake and the Finger Lakes region by the turn of the 21st century. Waterchestnut also migrated north into Lake Champlain through the Hudson-Champlain Canal and most likely reached Quebec through the Richelieu River system during the late 1990s. Waterchestnut was first found in Maryland in the late 1910s and reached the Potomac River during the early 1920s; widespread populations were present by the 1940s.

Description of the species

Waterchestnut is an ideal candidate for early detection programs because its appearance differs from all other plants found in North America and the species can often be identified early in its colonization cycle. Waterchestnut is an annual floating-leaved dicot that grows primarily in sluggish, shallow water. The habitat for this species includes lakes, ponds, reservoirs, sheltered margins of flowing water, freshwater wetlands and fresh to brackish estuaries. Waterchestnut usually grows in water less than 7 feet deep but has been found at depths of 12 to 15 feet. The species prefers thick, nutrient-rich organic sediments and an alkaline environment, but is tolerant of a wide pH range. Waterchestnut will not grow in salt water, although it can survive in brackish water with freshwater springs and groundwater input. The species grows aggressively and regularly produces as much as one pound of dry weight per square yard of surface area. Severe infestations can result in much greater biomass production; for example, waterchestnut populations growing in shallow impoundments in upstate New York have reportedly yielded almost 17,000 pounds of dry biomass per acre.

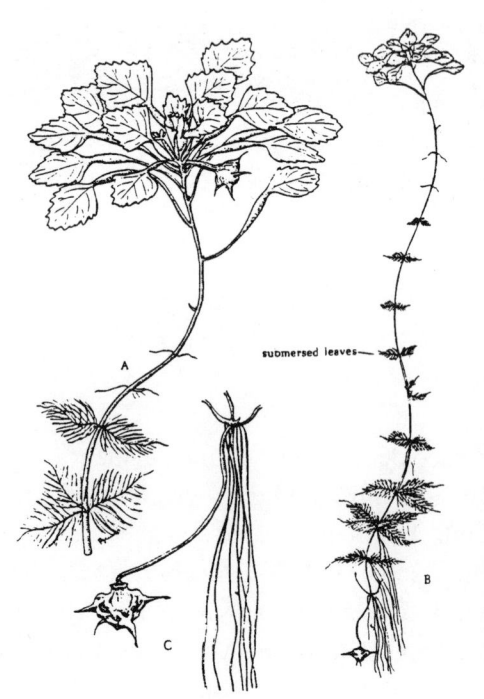

Submersed leaves of waterchestnut are pinnate (feather-like) and superficially resemble the finely dissected leaves of milfoils (*Myriophyllum* spp.). Submersed leaves are up to 4" long and are attached to the flexible stem in a whorl. Surface or floating leaves are palmate (divided like the fingers on a hand) and form a rosette of leaves that can be as broad as 1 foot in diameter. Leaf blades are 1 to 2" long and diamond shaped with a coarsely serrated (saw-toothed) margin. The upper sides of the leaves are bright green and the undersides are yellow-green with prominent veins. Rosettes form below the water surface and elongate to the surface by late spring – plants are buoyant due to inflated petioles or leafstalks (bladders) just below the rosette of leaves. Surface rosettes may initially be hidden within beds of other plants that produce floating leaves [e.g., watershield (*Brasenia* spp.), spatterdock (*Nuphar* spp.) and white waterlilies (*Nymphaea* spp.] and by smaller floating plants such as duckweed (*Lemna* spp.), watermeal (*Wolffia* spp.) and filamentous algae (Chapter 12). However, the prolific growth of waterchestnut will eventually create dense monocultures with as many as 50 rosettes per square yard and will crowd out desirable native plants. Beds of waterchestnut can be so extensive that they may completely cover the shallow zones of lakes and rivers and may obscure the margin between land and water.

Waterchestnut produces a single-seeded four-pronged nutlet with barbed spines. This structure is only produced by *Trapa natans* var. *natans* and allows for easy identification of the variety. The barbed spines are sharp enough to penetrate a wet suit – a painful experience for anyone unfortunate enough to step on one of these nutlets – and are the basis for the imaginative common names given to this plant. In addition to wreaking havoc on divers and swimmers, these nutlets figure prominently in the spread and propagation of this invasive species.

Reproduction
Many invasive species spread and reproduce from fragments, tubers, turions or underwater runners or stolons, but waterchestnut is an annual that reproduces solely from seeds. Small white flowers with yellow stamens are produced on the rosette after June, then drop into the water during summer and mature as nutlets between July and September. Each rosette produces 10 to 15 nutlets, which are capable of persisting for 10 to 15 years if kept moist in the sediment. Nutlets are around 1" wide, approximately 20% more dense than water and change from fleshy green to woody black by late summer. Mature nutlets drop from the plant and quickly sink into the sediment or wash to the shoreline, where the barbed spikes anchor the nutlet into the sediment. Parent plants disintegrate in the fall and seeds begin to germinate within a month after water temperatures warm to 50 °F or higher the following spring. A single nutlet can produce multiple rosettes because the rhizome can branch laterally to produce multiple upright stems.

Nutlets migrate between bodies of water by a variety of means. The most conspicuous vector for many years was humans, who intentionally introduced the waterchestnut as an ornamental. *Trapa natans* is listed as a federal "species of concern", but there are currently no explicit federal transport restrictions. Fortunately, a new appreciation of the environmental and economic problems that accompany establishment of this species and a network of state laws (including laws in NY, VT, NH, FL, MN and ME) that prohibit its transport have greatly reduced intentional introduction of waterchestnut. However, nutlets continue to move on currents between connected waterways, on the feathers, talons and webbed feet of numerous waterfowl and furred mammals, and especially on boat propellers, trailers and even foam bumpers on canoes.

Problems associated with waterchestnut
Infestations of waterchestnut cause problems similar to those of other invasive aquatic plants. Waterchestnut can form dense surface canopies that reduce sunlight penetration into the water column by 95% and crowd out other submersed and floating-leaved native plants and the fauna that rely on these plants for food and shelter. There is strong evidence that vallisneria or water celery (*Vallisneria americana*), a highly valued native plant, has been eliminated from many parts of the Hudson River after colonization by waterchestnut. This is due to the reduction in habitat available to vallisneria and to depletion of dissolved oxygen under large waterchestnut canopies, which also has a negative effect on small invertebrates. Large populations of waterchestnut create hostile environments for many desirable species such as banded killifish and spottail shiner and are often inhabited by

fauna that are more tolerant of adverse conditions, including rough fish species such as the common carp. Dense beds of waterchestnut can also entrap predatory birds seeking food within and underneath the surface canopy. Although waterchestnut canopies could potentially create significant pockets of still water to support mosquitoes, this has not been well documented in North American populations of waterchestnut.

Waterchestnut often grows under eutrophic conditions, in part because eutrophic bodies of water often create the thick organic sediments preferred by this plant and in part because waterchestnut grows in shallow waters where poor water clarity found in eutrophic waterways does not limit plant growth. Thick masses of leaves and stems generated by waterchestnut degrade and settle into the bottom sediments, which increases the organic content (and depth) of the sediment and contributes to greater turbidity and a cycle of increasing eutrophication. Bacterial degradation of this plant material can reduce dissolved oxygen, particularly at the end of the daily respiration cycle and when plants rapidly degrade in response to active management, such as herbicide treatment. Plant tissues also accumulate some heavy metals; this may occur with other highly abundant aquatic plants as well and may ultimately be a net benefit since these metals are removed from sediments or the water column.

Dense surface canopies of waterchestnut reduce water flow and impede boating and other forms of non-contact recreation, a particularly vexing problem since this plant often dominates navigable rivers and slow-moving water around marinas. Unlike submersed invasive plants and most floating-leaved plants, waterchestnut creates canopies that are impenetrable by even canoes and kayaks – the rosettes swallow paddles and significantly retard the momentum of the paddler. The same shallow waters frequented by canoers and kayakers are sometimes used for swimming, although the soft, thick organic sediments usually needed to support waterchestnut plants do not provide the ideal habitat for waders and swimmers. Waders willing to slog through dense populations of waterchestnut must carefully navigate through the nutlets commonly found along the shoreline and in the upper layer of near-shore sediments since stepping on the sharp barbs can cause deep puncture wounds. Dense mats create an additional safety concern – entanglement in waterchestnut beds may have contributed to drowning deaths in the Hudson River in 2001.

The most significant impact of waterchestnut infestations on humans may well be a reduction in aesthetics. Dense waterchestnut beds can completely cover the surface of shallow bodies of water and small ponds and will often carpet the near-shore areas of popular navigable rivers. The description grudgingly applied to waterhyacinth – "chokes out a water surface" – applies to waterchestnut as well.

Management options

During the past 100 years, many techniques have been used to manage waterchestnut. Unlike most invasive aquatic plants, waterchestnut has been effectively controlled and perhaps even eradicated in some bodies of water, but only after persistent effort. As with other invasive plants, best management of waterchestnut results from a vigilant prevention program. Weed watcher programs are particularly effective in controlling waterchestnut since the species is easily identified.

Once present in a body of water, waterchestnut can be controlled by physical and chemical techniques and may ultimately be managed by biological agents. Initial infestations, particularly when only a single rosette is found, can be pulled by hand (Chapter 6). The best window for removal of waterchestnut is from mid-June to mid-August – earlier efforts may result in regrowth or incomplete removal of nutlets, whereas later attempts might miss some nutlets or cause loosely attached seeds to dislodge. Plants should be flipped upside down immediately after removal to prevent dropping of seeds. Kayaks or canoes can be used for hand removal of waterchestnut; kayaks are more easily maneuvered through dense beds of waterchestnuts, but canoes carry more chestnut cargo. Hand removal programs led by cooperative extension offices, community groups, Boy Scout troops and volunteers have effectively controlled waterchestnut in Oneida Lake in central NY and in countless other smaller bodies of water throughout the Northeast.

Mechanical harvesting (Chapter 7) can effectively control large infestations of waterchestnut since the species is not spread by fragmentation, although cutting just the leaves (rosettes) from plants will likely leave nutlets in the system. Mechanical harvesting of plants after seeds have formed but before they mature can effectively break the reproductive cycle of the plant; however, the longevity and quantity of seeds in the sediment's seed bank may make it necessary to repeat the operation for at least 5 to 10 years to eradicate the species. A variety of state and federal agencies have used large mechanical harvesters to greatly reduce waterchestnut populations in Lake Champlain in Vermont and New York and in the Mohawk and Potomac Rivers. However, populations rapidly rebounded and returned to pre-harvesting densities when harvesting was suspended due to loss of funding.

Herbicides have also been used to control large-scale infestations of waterchestnut (Chapter 11). The herbicide used most often for control of this aquatic weed is 2,4–D, which is usually applied in early summer when plants are just reaching the water surface. Recently, triclopyr has also been used to control waterchestnut. Research is underway to determine whether glyphosate provides control of waterchestnut when applied directly to the rosette of surface leaves.

Grass carp (Chapter 10) have been used as biocontrol agents to manage waterchestnut in some bodies of water. However, grass carp are relatively indiscriminate feeders that find waterchestnut to be unpalatable, so few plants are consumed. Insect-based biocontrol (Chapter 9) may be a more promising alternative; researchers are currently evaluating a native leaf beetle (*Galerucella birmanica*) which has shown promise. However, this native beetle is a generalist feeder that consumes plants other than waterchestnut. Because successful biocontrol agents must be species-specific and feed only on a particular host plant (Chapter 8), this native beetle may not be a viable biocontrol option for waterchestnut.

Summary

Waterchestnut is one of the most invasive aquatic plants in the northeastern United States and has spread from its introduced range into neighboring states over the last 125 years. This species creates significant ecological damage, restricts human use of waterways and can be very difficult to control without consistent and persistent effort. However, waterchestnut is unique among invasive aquatic plants because it is easily detectable through citizen watch programs and can be controlled or even eradicated if caught early in its colonization. The species is an annual and can be managed by preventing seed production. Once established, waterchestnut requires significant resources to manage and vigilant use of mechanical or chemical control methods for 10 to 15 years to exhaust the reservoir of dormant seeds harbored in sediments.

For more information:
- Crow GE and CB Hellquist. 2000. Aquatic and wetland plants of northeastern North America. University of Wisconsin Press.
- Hummel M and E Kiviat. 2004. Review of world literature on water chestnut with implications for management in North America. Journal of Aquatic Plant Management 42:17-28. http://www.apms.org/japm/vol42/v42p17.pdf.
- Invasive plants of the eastern United States website. http://www.invasive.org/eastern/biocontrol/3WaterChestnut.html
- Kiviat E. 1993. Under the spreading water-chestnut. News from Hudsonia 9(1):1-6.
- Water chestnut management plan for central New York waterways website. http://74.125.47.132/search?q=cache:-iujZyAFklEJ:counties.cce.cornell.edu/onondaga/document/pdf/envi/Water%2520chestnut%2520Plan%2520w-o%2520Appendices.pdf+herbicid+water+chestnut+control&hl=en&ct=clnk&cd=13&gl=us

Photo and illustration credits:
Page 99: Waterchestnut infestation; The Nature Conservancy (photographer unknown)

Page 100: Line drawing; Barre Hellquist. From Crow GE and CB Hellquist. 1983. Aquatic vascular plants of New England: part 6. Trapaceae, Haloragaceae, Hippuridaceae. Station Bulletin 524. New Hampshire Agricultural Experiment Station, University of New Hampshire, Durham, NH.

Page 102: Waterchestnut plant; Hilary Smith, The Nature Conservancy

Linda S. Nelson
US Army Engineer Research & Development Center, Vicksburg MS
Linda.S.Nelson@usace.army.mil

Chapter 13.4: Giant and Common Salvinia

Salvinia molesta D.S. Mitchell; *Salvinia minima* Baker; free-floating ferns in the Salviniaceae family
Derived from *Salvinia* (after Antonio M. Salvini) and *molesta* (Latin: nuisance, annoying, troublesome) and *minima* (Latin: small, minor)

Introduced from Brazil (*Salvinia molesta*), Central and South America (*Salvinia minima*)
Found throughout the southern US

Introduction and spread

Water ferns in the genus *Salvinia* are members of the Salviniaceae family. There are 12 species of *Salvinia* reported worldwide, seven of which originate from the New World tropics. None of the *Salvinia* species are native to North America, but two species – *Salvinia minima* and *Salvinia molesta* – have been introduced and are currently established in the US. Both species were likely introduced into the US through the nursery trade as ornamental plants for water gardens or through the aquarium plant industry.

Salvinia molesta, commonly known as giant salvinia, is native to southeastern Brazil and was first found outside its native range in Sri Lanka in 1939. Giant salvinia quickly became a widespread weed problem in Sri Lanka, infesting rice paddies, reducing flows in irrigation channels and blocking navigation in transportation canals. Today, giant salvinia is considered one of the world's worst weeds and has become established in over 20 countries including Africa, India, Indonesia, Malaysia, Singapore, Papua New Guinea, Australia, New Zealand, Fiji, Cuba, Trinidad, Borneo, Columbia, Guyana, the Philippines and Puerto Rico.

The first report of giant salvinia outside of cultivation in the US occurred in 1995 when it was discovered in a small, private pond in South Carolina. Once identified, it was quickly eradicated from this site with the use of herbicides. Although this initial infestation was successfully eradicated, giant salvinia has since been reintroduced and has spread throughout the southern US. Significant infestations have been reported in more than 90 locations in 41 freshwater drainage areas of 12 states including Alabama, Arizona, California, Georgia, Hawaii, Florida, Louisiana, Mississippi, North and South Carolina, Texas and Virginia. Giant salvinia is currently listed as a Federal Noxious Weed by the US Department of Agriculture (www.aphis.usda.gov/ppq/weeds/), which prohibits its importation into the US as well as its transport across state lines. However, giant salvinia must be listed as a noxious species by individual states to prohibit sale and cultivation of the species within that state. Since it is not currently designated as a noxious weed by all states, the expansion of giant salvinia will likely continue across the US. Quarantine and sale of this plant by the nursery industry has been difficult to enforce nationwide. In fact, a recent survey of mail-order catalogs and on-line commercial vendors for water garden enthusiasts revealed that giant salvinia was among the many noxious aquatic plants readily available for purchase over the internet.

Salvinia minima, hereafter referred to as common salvinia, is native to Central and South America. Outside its native range it has established in Bermuda, Puerto Rico, Spain and North America. Common salvinia was first reported in the US in 1928 along the St. John's River in Florida. The source of this first introduction to a natural area was likely the result of an unintentional release from a grower whose cultivation ponds had flooded. Since then, populations have been recorded in more than 80 freshwater drainage areas across southern and southeastern states including Texas, Arkansas, Louisiana, Mississippi, Alabama, Florida, North and South Carolina and Georgia. Similar to giant salvinia, common salvinia is widely available through the water garden trade. Although it continues to infest new regions, common salvinia is not listed as a Federal Noxious Weed; however, it is currently listed as a prohibited plant in the state of Texas.

Description of the species

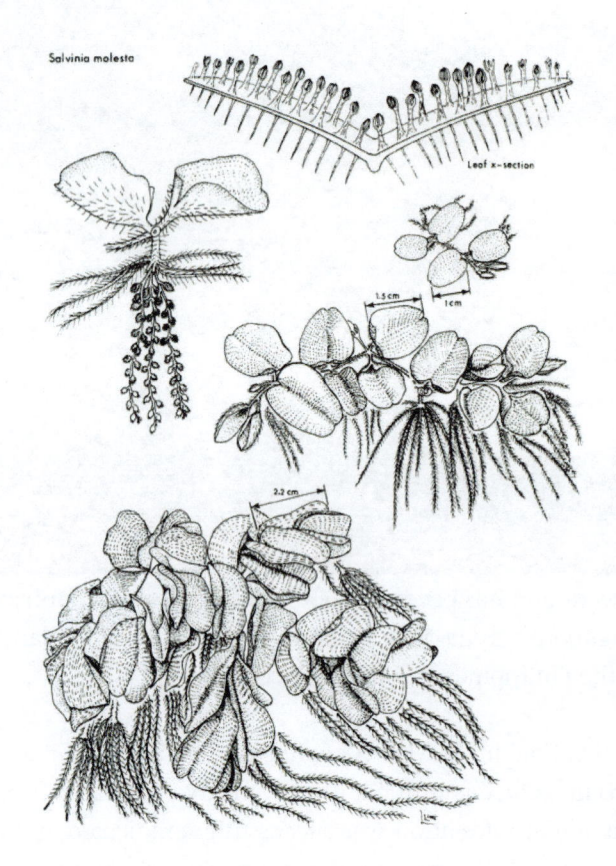

Common and giant salvinia are free-floating aquatic ferns with a horizontal stem or rhizome that floats at or just below the water surface. A pair of floating leaves or fronds (leaves of ferns are referred to as "fronds") are produced at each node along the rhizome. Fronds are bright green in color, oval in shape, possess a central midrib and are covered with numerous stiff, white hairs. It is thought that the function of these leaf hairs is to repel water and thus aid in plant buoyancy. An easy way to distinguish giant salvinia from common salvinia is by the shape of the hairs on the upper surface of floating fronds. The hairs on the fronds of giant salvinia form cage-like structures at the tip that resemble an eggbeater or kitchen whisk, whereas the hairs on common salvinia fronds are open at the tip and have a fringed appearance (see next page).

Common and giant salvinia lack true "roots" but possess delicate, finely-dissected submersed fronds. Submersed fronds are brown and resemble roots and serve a similar function by absorbing nutrients from the water. Sporocarps (structures that hold the fern's spores) are borne in chains or clusters on submersed stalks but do not bear fertile spores. Sporocarps are not found at all plant nodes but often develop and are more abundant later in the growing season or when nutrient conditions are poor.

Both giant and common salvinia favor stagnant or slow-moving water habitats of lakes, ponds, rivers, streams, oxbows, ditches, canals, swamps, marshes and rice fields. Under favorable growing conditions, both species can form dense, expansive plant mats that can completely cover the water surface. Optimal growing conditions include full sunlight and warm (75 to 85 °F), nutrient-rich waters with a pH of 6 to 7.5. Upper and lower temperature thresholds for growth are about 95 and 50 °F, respectively. Both giant and common salvinia have a low tolerance to salinity and cannot survive in brackish or marine environments.

Common salvinia (upper photo)

Giant salvinia (lower photo)

Reproduction

Giant and common salvinia are ferns, so they do not produce flowers or seed. As mentioned above, both species produce sporocarps that may contain spores but the spores are not viable. As a result, giant and common salvinia reproduce solely by vegetative means through fragmentation or the production of new plants from lateral and terminal buds. Stems may have as many as 5 buds per node and each bud is capable of developing new fronds. In addition, horizontal stems or rhizomes break apart very easily and produce fragments that disperse and develop into mature individual plants.

An individual giant salvinia can double in size in as little as 5 to 7 days when conditions are favorable. Some reports have calculated that a single giant salvinia plant can multiply to cover 40 square miles in 3 months under optimal growing conditions. With such an explosive growth rate, giant salvinia can quickly cover lakes and rivers, forming vegetative mats up to 3 feet thick. Common salvinia also has a rapid growth rate and can form dense mats, but is often less aggressive than giant salvinia.

The major means of dispersal within and among lakes for giant and common salvinia is vegetative spread by fragmentation. Plant populations expand laterally within a lake through rhizome and lateral bud growth, whereas long distance dispersal is mostly the result of fragmentation. Plants easily adhere to boats, trailers, motors and other amphibious vehicles and can be transported to new locations. Animals (livestock, turtles, wading birds and waterfowl) may also contribute to the spread and dispersal of salvinia.

Problems associated with giant and common salvinia

Both giant and common salvinia can alter aquatic ecosystems in many ways. Dense growths can form a physical barrier on the water surface and hinder recreational activities such as boating, swimming, fishing and water skiing. Vegetative mats of salvinia can also impede navigation, impair flood control, limit irrigation, clog water intakes, decrease waterfront property values and cause problems in rice, catfish and crawfish production systems. Occasionally, other plant species (including grasses and small trees) will colonize mats of giant salvinia and create massive floating islands that can trap sediments and cause waterbodies to fill in over time.

Ecologically, extensive salvinia mats can restrict light penetration and impede gas exchange between the water and atmosphere. Limiting light availability reduces photosynthesis of submersed aquatic plant communities and reduces water temperature. Low dissolved oxygen levels in the water are detrimental to fishes and other aquatic organisms and promote the accumulation of organic matter as microbial degradation is reduced. Changes in water quality can significantly impact the health of aquatic habitats and often result in declines in number and diversity of plant, invertebrate and animal communities. The loss of open water habitat also reduces the use of these areas by migrating waterfowl and wading birds (Chapters 3 and 4).

Public health issues are also of concern. Both species of salvinia provide breeding habitats for mosquitoes and associated mosquito-born illnesses (e.g., West Nile virus, malaria, encephalitis—Chapter 5). In Sri Lanka, it was reported that giant salvinia served as an important host plant and breeding habitat for mosquitoes which transmit filariasis (elephantiasis). Increases in the occurrence of schistosomiasis have also been linked with large infestations of giant salvinia in developing countries.

Management options

Giant and common salvinia can be managed using herbicides, biocontrol agents, manual or mechanical harvesting, water level manipulation or a combination of these methods. Selecting the best management strategy depends on site-specific management goals and objectives, site characteristics, size and density of the infestation, proximity to sensitive plant or animal species, water body uses and budget constraints. The key to successfully managing giant and common salvinia is to recognize the problem early when infestations are small and can be easily contained. Once giant or common salvinia become well established and cover large areas, management becomes more difficult, time consuming and costly and may require multiple applications of a treatment method over a number of years to achieve maintenance control.

Chemical control (Chapter 11). Herbicides can provide effective short and/or long-term control of giant and common salvinia depending on the choice of product and method of application. Of the herbicides currently registered by the US Environmental Protection Agency for use in aquatic sites, six provide excellent control (> 90%) of giant or common salvinia. The most widely used herbicides against these weed species include diquat, glyphosate and carfentrazone-ethyl. Diquat and carfentrazone-ethyl are non-selective contact herbicides that are typically applied as foliar sprays. Injury symptoms (severe leaf browning) are visible one day following application and plant death occurs within 3 to 4 days of treatment. Contact herbicides are fast-acting but have little or no movement inside plant tissues, so only plant tissues that come into contact with the herbicide are affected. Glyphosate is a non-selective, systemic herbicide that is applied to foliage, absorbed through the leaves and moves throughout the plant. Injury symptoms (leaf yellowing and browning) appear seven days after glyphosate application and plant death occurs by 28 days after treatment.

Other systemic herbicides that are effective, but slower-acting and used to a lesser extent against these two salvinia species, include imazamox, fluridone and penoxsulam. Imazamox is effective on common salvinia but shows little or no activity on giant salvinia. Both species are susceptible to penoxsulam and fluridone. Fluridone and penoxsulam require long contact times (60 to 90 days) to achieve control of salvinia, whereas imazamox has a shorter contact time requirement (7 days). Contact time refers to the length of time the target plant must be in contact with or exposed to a lethal dose of herbicide to achieve control. If contact time is not maintained because of water exchange or other factors that can cause dilution, plant control will be reduced. Imazamox and penoxsulam can be applied as a foliar spray or as a submersed application to the water column, whereas fluridone is effective only as an in-water treatment. Although in-water herbicide applications can be effective for treating these floating weed species, this method may not be feasible for sites where high water exchange or flow affect herbicide contact time and may be prohibitively expensive in larger systems.

Giant and common salvinia can be difficult to manage using herbicides because they are small floating plants that produce dense stands with plants layered on top of one another. This layering of plants presents a challenge when applying herbicides because plants in lower layers of the mats are protected from herbicides by plants in the upper layers of the mats. If plants are dense and a thick vegetative mat has formed, multiple applications will be required to achieve successful long-term control. In addition, giant and common salvinia can survive short dewatering or drawdown events and can persist on moist soils; therefore, spraying shoreline areas in addition to plants on the water surface is important to prevent reinfestation via surviving plant material. Long-term management with herbicides requires follow-up monitoring to spot-spray any plant material that survived the initial application. As a good management practice, herbicides should be routinely rotated and/or combined with other control strategies to minimize the potential development of herbicide resistance.

Biological control (Chapters 8 and 9). Several insects have been investigated as biological control agents against salvinia species, but the salvinia weevil (*Cyrtobagous salviniae*) is recognized throughout the world as the insect of choice for management of giant and common salvinia. This insect feeds and reproduces only on plants in the Salviniaceae family. The salvinia weevil is a small (less than 1/16" long) black weevil native to South America. Adults feed on floating fronds and rhizomes but prefer newly formed buds. The larvae of the salvinia weevil are white, 1/8" long and feed within the floating and submersed fronds, rhizomes and buds. Feeding by the larvae is often more destructive than that of adults. The combined feeding action of adults and larvae can be devastating and can impact field populations of giant and common salvinia in several months as opposed to the longer periods of time required by other insect biocontrol agents. Attacked plants turn brown in small patches that merge together until the whole colony loses structural integrity, becomes waterlogged and sinks.

Although never intentionally released, the salvinia weevil was first detected in Florida in 1960, where it is now widespread and feeds primarily on common salvinia. Initial attempts to release weevils collected from Florida to manage giant salvinia in Texas and Louisiana were ineffective. This prompted researchers to seek permission from the Technical Advisory Group and the USDA-APHIS-PPQ (US Department of Agriculture, Animal and Plant Health Inspection Service, Plant Protection and Quarantine – see Chapter 8), to release a strain of the salvinia weevil from Australia which was highly effective in overseas applications. Permission was granted in 2001 and the Australian weevils were released in east Texas and western Louisiana only. The weevils have since become established and are beginning to impact giant salvinia in these localized release sites.

Herbivorous fish such as triploid grass carp (Chapter 10) and tilapia (*Oreochromis* sp.) have been evaluated as possible biocontrol agents against salvinia with limited success. Laboratory feeding studies showed that while tilapia will consume giant salvinia, it is not their preferred food if other food sources are available. Other studies have shown that salvinia provides little nutritional benefit to herbivorous fishes.

Mechanical control (Chapter 7). The effectiveness of mechanical methods or manual removal is limited but may be useful in the early stages of an infestation or when a localized population is found on a small water body. If mechanical harvesting methods are employed, plant material must be properly disposed of in upland areas where the potential for contamination of other water bodies is minimized. Mechanical removal is not economically feasible once giant or common salvinia is well established and covers large areas. However, combining mechanical removal with herbicide applications can be an effective integrated weed management strategy. For example, in 2003, the Hawaii Department of Agriculture was successful in controlling 300 acres of giant salvinia on Lake Wilson on Oahu using multiple applications of the herbicide glyphosate combined with mechanical removal techniques. Excavated plant material was safely disposed of in nearby pineapple fields.

Other management options (Chapter 6). Floating booms have been used to contain and limit the spread of giant and common salvinia in some systems but are generally only utilized to confine plants to one location while other management strategies such as herbicides or weevils are deployed. Drawdowns can be a low-cost, effective management approach in some situations where water levels can be manipulated. However, dewatering must occur over a long period of time to allow plants to become stranded on dry land where they will desiccate and/or be exposed to freezing temperatures.

Plant material can remain viable for several months if stranded shoreline mats are dense and underlying moisture is present. Decaying plant material along shorelines can be unsightly and plant fragments can easily be blown back into the system.

Summary

Giant and common salvinia are fast-growing, mat-forming aquatic ferns that can quickly cover the water surface of lakes, rivers and other wetland habitats. They are aggressive competitors that reproduce only by vegetative means. The plants can tolerate a wide range of growing conditions but prefer warm, nutrient-rich waters and full sunlight. Giant and common salvinia prefer freshwater environments and will not colonize saline or brackish waters. Once established, herbicides can be used to effectively manage these plants; however, multiple applications, follow-up monitoring and spot treatments may be required to maintain long-term control. Introducing insect biocontrol agents such as the salvinia weevil can be effective for maintenance control in some systems. The salvinia weevil has been especially successful in Florida for keeping common salvinia populations in check. Preventing the spread of this plant through citizen watch programs, boat launch surveillance and enforcement and compliance with laws to prevent the cultivation, sale and transport of these species will be important for containing and minimizing further spread of giant and common salvinia in the US.

For more information:

•Holm LG, DL Plucknett, JV Pancho and JP Herberger. 1977. The world's worst weeds: distribution and biology. University Press of Hawaii.
•McFarland DG, LS Nelson, MJ Grodowitz, RM Smart and CS Owens. 2004. *Salvinia molesta* D.S. Mitchell (giant salvinia) in the United States: a review of species ecology and approaches to management. Aquatic Plant Control Research Program ERDC/EL SR-04-2.
http://el.erdc.usace.army.mil/elpubs/pdf/srel04-2.pdf
•McIntosh D, C King and K Fitzsimmons. 2003. Tilapia for biological control of giant salvinia. Journal of Aquatic Plant Management 41:28-31. http://www.apms.org/japm/vol41/v41p28.pdf
•Oliver JD. 1993. A review of the biology of giant salvinia (*Salvinia molesta* Mitchell). Journal of Aquatic Plant Management 31:227-231. http://www.apms.org/japm/vol31/v31p227.pdf
•Websites with information on giant and common salvinia:
http://salvinia.er.usgs.gov/html/identification1.html
http://salvinia.er.usgs.gov/index.html

Photo and illustration credits:

Page 105: Giant salvinia at Lake Wilson, Oahu; Linda Nelson, USACE ERDC
Page 106: Line drawing; University of Florida Center for Aquatic and Invasive Plants
Page 107 upper: Common salvinia; Ted Center, bugwood.org
Page 108 lower: Giant salvinia; Mic Julien, bugwood.org
Page 110: *Cyrtobagous salviniae* on giant salvinia frond; Scott Bauer, bugwood.org

Lyn A. Gettys
University of Florida, Gainesville FL
lgettys@ufl.edu

Chapter 13.5: Waterhyacinth

Eichhornia crassipes (Mart.) Solms; floating plant in the Pontederiaceae (pickerelweed) family
Derived from *Eichhorn* [Johann Albrecht Friedrich Eichhorn (1779-1856), Prussian minister of education and public welfare] and *crass* (Latin: thick)
"plant with thick leaf stalks"

Introduced from Brazil to New Orleans in 1884
Present throughout the southeastern US and California, Hawaii and the Caribbean area

Introduction and spread

Eichhornia crassipes is one of around seven species in the genus *Eichhornia*, all of which are native to South America. Waterhyacinth is native to the Amazon River and has been widely introduced throughout the tropical regions of the world, most recently occurring in Lake Victoria in East Africa. The first known introduction of waterhyacinth to North America was at the Cotton States Exposition in New Orleans in 1884. The species was initially cultivated as an ornamental but quickly escaped cultivation and invaded other parts of the southeastern US. Waterhyacinth must have been a botanical curiosity due to its size, floating growth habit and the beauty of its very short-lived purple flower spikes. Mr. Fuller (the owner of Edgewater Grove, 7 miles upstream of Palatka on the St. Johns River) introduced this "beauty" to Florida around 1890. It was initially grown in Mr. Fuller's fountain pond and excess growth was cast into the St. Johns River, where within a short time it covered the half-mile wide river from bank to bank at several locations. Waterhyacinth spreads very rapidly; for example, the species covered 126,000 acres of Florida's surface water within 70 years of its arrival in that state. Waterhyacinth is present throughout the southeastern US, California, Hawaii and the Virgin Islands, but is considered eradicated in Arizona, Arkansas and Washington State. Populations of waterhyacinth have been reported in other states, including New York, Kentucky, Tennessee and Missouri and plants are intentionally introduced to farm fish ponds in southern Arizona and southern Delaware. This species is not cold-hardy and has not established permanent populations in more temperate areas outside the southern US. Waterhyacinth will survive moderate freezes but requires temperatures of greater than

50 °F to produce new growth. A number of states, including Florida, South Carolina and Puerto Rico, prohibit the sale of waterhyacinth, but the species is still available for purchase from aquarium supply stores, aquatic plant nurseries and internet sources in other states. Waterhyacinth spreads in natural systems by producing seedlings and daughter rosettes – small plantlets that are attached to the mother plant by a floating stolon or runner. Rosettes can easily become caught in boat trailers or live wells, which results in the introduction of the species to new bodies of water. Waterhyacinth is also spread by uninformed water garden and pond owners, who (along with Mr. Fuller in the 1890s) believe they are beautifying canals and lakes by tossing extra plants into natural systems.

Description of the species

Waterhyacinth is a floating flowering monocot that grows as an annual (in temperate regions) or as a perennial (in tropical and subtropical climates) in all types of bodies of water. Muddy or turbid water often limits growth of submersed plants, but because waterhyacinth is a floating plant, it is unaffected by these conditions. The leaves of waterhyacinth are thick, glossy, waterproof and rounded with a heart-shaped base. Each leaf can reach up to three feet in length and is borne singly on a spongy, inflated petiole (leaf stalk). Leaves are attached to one another at the base of the petiole to form a rosette that is free-floating, although plants will sometimes root in soft saturated sediments when stranded by drought or wave action. The dark purple to black roots of waterhyacinth are long and feathery and hang beneath the rosette of leaves. Waterhyacinth grows throughout the year in the tropics, but freezing temperatures kill the leaves of the plant in the northern portions of its range. Cold-damaged leaves then fold down and protect the meristem, which grows at or immediately below the surface of the water.

The most striking feature of waterhyacinth is the spike of large, showy flowers produced from the center of the rosette of leaves. Flowers are borne in groups of 8 to 15 on a single spike that can rise up to 20" above the rosette. Each flower is up to 3" tall and has six lavender-blue to purple petals, with the uppermost petal marked by a yellow "eye-spot." Flowers are short-lived, with each lasting only one or two days, but a spike may be showy for up to a week since only a few flowers open each day. Flowering is indeterminate – flowers at the base of the spike open first and flowers at the top of the spike open last. After flowers are fertilized, the spike bends and dips into the water, where many tiny seeds are produced in capsules. Mature seeds fall to the bottom of the body of water, where they remain dormant until sediments are exposed after water levels fall due to drought.

Waterhyacinth is sometimes confused with native frog's-bit (*Limnobium spongia*), because both are floating plants with rounded leaves borne in rosettes. However, the roots of waterhyacinth are black and feathery, whereas the roots of frog's-bit are thicker and white. In addition, the petioles of frog's-bit are usually slender, while the petioles of waterhyacinth are often spongy and bladder-like. Finally, flowers of frog's-bit are small, white and much less showy than those of waterhyacinth.

Reproduction
Waterhyacinth spreads by both seed and vegetative reproduction. As noted above, seeds are tiny and remain dormant until conditions are favorable for germination. Some reports suggest that seeds germinate best after they have dried and others say that seeds must be exposed to alternating warm and cold temperatures before they will germinate. Seed reproduction can be important in temperate climates since waterhyacinth is killed by freezing temperatures and recolonization in spring may be dependent on the seed bank established during the previous growing seasons. Once seeds have germinated and conditions are favorable for growth, waterhyacinth rapidly produces new daughter plants, or ramets, from horizontally growing stolons. Daughter plants can be produced in as little as 5 days under optimal growing conditions and populations can double in size in as little as 6 to 18 days, so the rapid growth and spread of waterhyacinth is due primarily to this type of vegetative reproduction.

Problems associated with waterhyacinth
Waterhyacinth grows almost entirely on the surface of the water as a floating plant and its growth potential is limited only by temperature and the availability of nutrients. Waterhyacinth prefers an environment similar to that favored by desirable fish populations – mesotrophic and eutrophic habitats with an adequate supply of calcium and a pH ranging from 6.5 to 9.5. There is no doubt that waterhyacinth is a serious aquatic weed. Under optimum conditions, an undisturbed population of waterhyacinth is composed of about 10 plants per square foot and has a fresh weight of 10 pounds. An acre (43,560 square feet) of waterhyacinth would therefore be home to about 435,600 plants with

a fresh weight of around 200 tons. Since 95% of the plant weight is attributable to water, only 5% of the fresh weight – about 10 tons per acre – remains after plants are harvested and dried.

Waterhyacinth may not be as productive as most agricultural crops; however, trying to remove or stop 200 tons of live waterhyacinths from jamming against a bridge or clogging a waterway is no simple task! Large colonies of linked mother and daughter plants form dense rafts or mats that can quickly cover a body of water from shore to shore. Left undisturbed, floating mats of waterhyacinth provide a perfect substrate or "island" to support the growth of additional grasses, herbaceous plants and even small trees, which further bind the floating mat together. These mats interfere with human use of waters. For example, large populations of waterhyacinth can restrict recreational and commercial activities and can make boating, fishing and swimming impossible. In addition, water flow is greatly reduced where mats of waterhyacinth are present, which can impede irrigation and flood control efforts. Infestations of waterhyacinth can have serious ecological impacts as well. Dense waterhyacinth populations also reduce species richness or plant diversity by limiting light availability to native submersed plants and by crushing communities of emergent plants along the shoreline. The loss of these plants also eliminates habitats for animals that depend on native plants for shelter, nesting and food. In addition, large mats block the air-water interface and reduce dissolved oxygen, which makes the system uninhabitable to fish and other aquatic fauna.

Management options

The best method to control waterhyacinth is to prevent the species from entering a water body. The sale and interstate shipment of a closely related species [rooted waterhyacinth (*E. azurea*)] is prohibited by the Federal Noxious Weed List and its introduction into the US has been avoided thus far. Waterhyacinth (*E. crassipes*) is not on the Federal Noxious Weeds List because the species was already widely distributed in the US at the time the Federal Noxious Weed Acts were developed. In spite of these prohibitions, waterhyacinth still manages to slowly increase its range and to colonize new bodies of water.

Physical (Chapter 6) or mechanical (Chapter 7) control measures such as hand removal or mechanical harvesters should be designed to prevent the spread of waterhyacinth plantlets to other parts of the water body. Hand removal is labor-intensive and typically involves raking plants to the shoreline or into a boat. This very laborious task can seem deceptively easy; a pond that is a single acre in size may look small, but can host up to 200 tons of waterhyacinth that must be pulled out by hand! Plants are then offloaded along the shoreline until they desiccate and die. Hand removal may be an effective means to control waterhyacinth in small ponds, but is not practical in larger systems. Mechanical harvesting is usually used to remove plants from larger systems and involves heavy machinery that ranges from a backhoe on a barge to specialized equipment. A problem associated with mechanical harvesting of waterhyacinth is disposal of the harvested plants. Waterhyacinth vegetation has been used to make furniture, baskets and other items in some parts of the world and has been evaluated for its potential as mulch, cattle feed, biofuel production and other uses, but its utility is very limited. As a result, most harvested waterhyacinth is generally disposed of in farm fields or a landfill. Hand removal of waterhyacinth from ponds is best employed after herbicide application has been used to control the majority of the plants. Regular removal of missed plants and any plants growing from seeds after herbicide treatment will prevent waterhyacinth from reinfesting the pond.

Drawdowns can be used to "strand" and desiccate waterhyacinth on exposed shorelines, but the time required to effectively dry large mats of plants can be long. Also, drawdowns and drought have been known to trigger seed germination and plants reestablish quickly when water levels rise. Therefore, most waterhyacinth management programs in the US rely on the use of herbicides in conjunction with established insect biocontrol agents. Waterhyacinth weevils (*Neochetina* spp.) were introduced and established in the early 1970s (Chapter 9). The weevils are found throughout the range of waterhyacinth but in most areas the insects only slow plant growth and reproduction and do not provide adequate control of the weed. As a result, herbicides are used in maintenance control programs to keep plant populations low and to reduce growth potential of waterhyacinth. Herbicide selection is based on water use, selectivity to reduce damage to nontarget native plants and cost (Chapter 11). Several herbicides are commonly used as foliar sprays to selectively control waterhyacinth. Contact herbicides – including diquat, copper and endothall – are quickly absorbed by plant tissue and are fast-acting, whereas systemic herbicides – including 2,4–D, glyphosate, imazamox and penoxsulam – provide slower but effective control.

Summary
Waterhyacinth is one of the world's worst aquatic weeds and causes problems in all tropical and subtropical continents. Its current distribution in the US is primarily from East Central Texas to the Atlantic Coast and north to coastal North Carolina. It also occurs in the Sacramento River Delta in California. Although waterhyacinth is occasionally found north of the central US, the species typically does not persist where waterways are subject to ice formation and prolonged freezing temperatures. Florida and the Gulf states are particularly impacted by waterhyacinth due to the moderate climate and shallow, naturally nutrient-rich lakes, but the species can colonize virtually any region in North America where winter temperatures remain above freezing and mesotrophic or eutrophic waters are present. Aggressive maintenance control programs have kept populations of waterhyacinth in check in most areas, but these efforts must be employed on a continual basis to avoid population explosions of this noxious invasive species.

For more information:
•Anonymous. 2008 (access date). Flora of North America: Eichhornia crassipes. FNA 26:39-41. http://www.efloras.org/florataxon.aspx?flora_id=1&taxon_id=200027394
•Buker GE. 1982. Engineers vs. Florida's green menace. The Florida Historical Quarterly. April 1982, pp. 413-427.
•Cook CDK, BJ Gut, EM Rix, J Schneller and M Seitz. 1974. Water plants of the world: a manual for the identification of the genera of freshwater macrophytes. Dr. W Junk b.v., Publishers, The Hague.
•Gopal B. 1987. Aquatic plant studies 1: water hyacinth. Elsevier, Amsterdam, The Netherlands.

Photo and illustration credits
Page 113: Waterhyacinth infestation; University of Florida Center for Aquatic and Invasive Plants (photographer unknown)
Page 114: Line drawing; University of Florida Center for Aquatic and Invasive Plants
Page 115: Waterhyacinth; University of Florida Center for Aquatic and Invasive Plants (photographer unknown)

Chapter 13.5

Robert L. Johnson
Cornell University, Ithaca NY
rlj5@cornell.edu

Chapter 13.6: Purple Loosestrife

Lythrum salicaria L.; erect, emergent perennial herb in the Lythraceae (loosestrife) family
Derived from *lythrum* (Greek: blood) and *salicaria* (Latin: willow-like)
"plant that stops blood and is willow-like"

Introduced from Europe to the east coast of North America in the early 1800s
Present in every state throughout the US except for Florida, and found in all Canadian provinces

Introduction and spread

Lythrum salicaria L. (purple loosestrife) is often referred to as "the purple plague" in North America and is native to Europe and Asia. Purple loosestrife is an aggressive invasive plant that was deliberately introduced to the eastern coast of North America in the early 1800s. Settlers of the region valued the plant as an ornamental for perennial gardens and used the species as a medicinal herb to treat dysentery, diarrhea, bleeding and ulcers. The honey trade also increased regional seed propagation of the plant because it was favored as bee forage. In addition, European ships contributed to the spread of purple loosestrife by releasing ballast water and delivering shipments of wool that contained seeds of the species. By the 1830s, purple loosestrife had become established along the New England seaboard and the range of the species further expanded throughout New York State and the St. Lawrence River Valley through inland canals constructed in the late 1880s. As road systems expanded and commercial distribution of the plant by the nursery trade increased, purple loosestrife spread westward and southward and can now be found in every state and province of the US and Canada, except for Florida. Purple loosestrife grows in most freshwater wetlands but also tolerates a wide range of environmental conditions and can spread to both tidal and non-tidal brackish waters.

Description of the species

Purple loosestrife is an erect, emergent perennial dicot herb with a dense, bushy appearance. The species tolerates a wide range of wetland environments and grows in habitats ranging from pastures with moist soil to sites with shallow water such as marshes and lakeshores. Established plants can tolerate a variety of soil conditions, including soils that are dry or permanently flooded and soils that are low in nutrients and pH. In addition, plants can grow in rock crevasses, on gravel, sand, clay or organic soils. Purple loosestrife can grow from four to ten feet in height and has a dense canopy of stems that emerge from its wide-topped crown. Each plant produces as many as 50 square, hard, red to purple stems that arise from a single root mass. Leaves are 1-1/2 to 4" long and 2/10 to 6/10" wide and are lance-shaped, stalk-less, heart-shaped or rounded at the base and borne in an opposite or whorled arrangement. Purple loosestrife produces flowers with magenta, purple, pink or white petals that are 4/10 to 8/10" long. The species blooms throughout most of the summer, which adds to its appeal as an ornamental plant and as a favorite of beekeepers. The reddish-brown seeds are very small (1/25" long) and are often produced during the first growing season. Purple loosestrife is often confused with a number of plants with spikes of purple flowers, including gayfeather (*Liatris pycnostachya*), blue vervain (*Verbena hastata*) and fireweed (*Epilobium angustifolium*). However, the species most closely resembles the native winged loosestrife (*Lythrum alatum*) and *Lythrum virgatum* L., a nonnative cultivated purple loosestrife. *L. virgatum* is very similar to purple loosestrife in appearance and was formerly classified as a separate species, but is now considered by some to be a subspecies or variant form of purple loosestrife.

Reproduction

The extended flowering season of purple loosestrife typically lasts from June to September and allows each plant to produce as many as 3 million seeds each year. Long-tongued insects, including bees and butterflies, serve as pollinators. Seeds are dispersed by water and can "hitchhike" in mud that adheres to wildlife, livestock and people. Seed survival can be as high as 60 to 70%, which produces a sizeable seedbank in only a few years. Germination occurs in open, wet soils as temperatures increase in the spring, but seeds can remain dormant and viable for many years in the soil. In addition, submersed seeds can survive for up to 20 months in flooded conditions. Purple loosestrife readily colonizes newly disturbed areas because of its high production of viable seeds with multiple modes of dispersal. Disturbed areas with exposed soil are most vulnerable to invasion and rapid colonization by purple loosestrife because these sites provide ideal conditions for seed germination and usually lack native plants that compete with the weed for resources. Purple loosestrife spreads predominately via seed dispersal, but can also spread vegetatively by producing new shoots and roots from clipped, trampled or buried plants. Purple loosestrife's ability to reproduce via vegetative means is especially important when adopting management strategies because mechanical or physical control efforts can inadvertently spread harvested plant fragments and create new infestation sites. In addition, disturbances in the form of changes in water levels from drought or a planned water drawdown provide ideal conditions for maximum seed germination and growth.

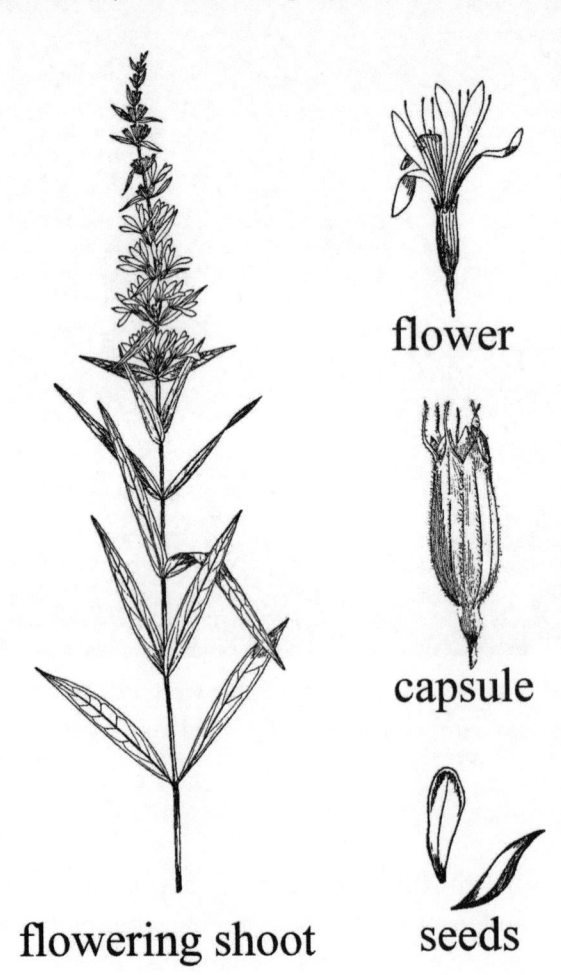

Problems associated with purple loosestrife

Purple loosestrife aggressively invades many types of wetlands, including freshwater wet meadows, tidal and non-tidal marshes, river and stream banks, pond edges, reservoirs and roadside ditches. The formation of dense, monotypic stands of purple loosestrife suppresses native plant species, decreases biodiversity and leads to a change in the wetland's community structure and hydrological functioning, while eliminating open water habitat in many locations. Around 200,000 acres of wetlands are lost in the US every year due to invasions of purple loosestrife and as much as 45 to $50 million per year is spent on efforts to control the growth of this species. In addition to funds spent on control efforts, economic losses to agriculture can exceed millions of dollars annually when purple loosestrife invades irrigation systems. Also, entire crops of wild rice may be lost when this species invades shallow lakes and bays dominated by wild rice, which results in great economic loss to agricultural communities.

Purple loosestrife alters the physical makeup of a wetland, but the species can change the chemical properties of the wetland as well. For example, leaves of purple loosestrife decompose rapidly after being shed in the fall and the nutrients released during decomposition are quickly flushed out of the wetland. In contrast, the vegetation of native species does not fully decompose until the following spring and nutrients are maintained in the wetland throughout the fall and winter. This difference in the timing of nutrient release means that wetland decomposers have fewer nutrients available to subsidize peak population growth in the spring, which alters the structure of the food web. The effects of altered water chemistry extend to many fauna in aquatic ecosystems as well. For example, chemicals released during the decomposition of purple loosestrife leaves can slow the development of certain frog tadpoles, which decreases the frog's chance of surviving its first winter. Recent research at Cornell University suggests that threats to amphibians by nonnative plants may be underestimated. Their data indicate that organisms that breathe through gills (especially *Bufo americanus*, the American toad) are sensitive to the high concentration of tannins naturally produced during purple loosestrife decomposition.

Purple loosestrife further affects the wildlife communities of wetlands through a variety of other means. The species is a very poor food source for herbivores and crowds out species that are more beneficial to the wetland food web. As a result, stands of purple loosestrife can jeopardize threatened and endangered plants and wildlife, especially in the northern US. For example, the bog turtle has lost extensive basking and breeding habitat due to the introduction of this aggressive plant. Purple loosestrife also displaces native plants such as cattail and bulrush, which provide high quality habitat to numerous nesting birds and aquatic furbearers. Wetland specialists such as the marsh wren or least bittern (Chapter 4) prefer sturdy nesting sites such as cattail-dominated wetlands and are unable to utilize purple loosestrife for their nests. Also, muskrat, beaver and waterfowl prefer cattail marshes and are more able to utilize these sites that are dominated by native plants as compared to dense, monotypic populations of purple loosestrife.

A primary problem associated with purple loosestrife is its attractiveness. European immigrants to the US deliberately imported purple loosestrife as an ornamental plant in the 1800s and homeowners still actively plant the species today. Purple loosestrife may add a welcome burst of color to an otherwise dull private garden or pond, but the adaptability and aggressiveness of this plant can quickly wreak havoc on the unsuspecting homeowner's backyard. The sale or distribution of purple loosestrife is illegal in many states; however, nurseries and greenhouses sell the plant in many areas across the country and it continues to be included in some seed mixes. Consumers should always read seed package labels before purchasing in order to ensure that this aggressive nonnative plant is not included in the mix.

Management options

The best way to stop an invasion of purple loosestrife is to be aware of pioneering plants and small isolated colonies. In these cases, hand removal of small, isolated stands is an effective preventative control method. The use of physical (Chapter 6) and mechanical (Chapter 7) control methods may provide annual control of low-density invasions and can include water level manipulation, hand removal, cutting and burning. When using these methods, treatment must be completed before seeds are produced to avoid seed dispersal and contributions to the seed bank. It is also essential to remove roots from the soil since plants will regrow from broken roots or root fragments. Removal of flowering spikes will prevent seed formation and cutting or harvesting stems at the ground level will inhibit growth temporarily. While these methods temporarily halt growth, they should be used in conjunction with herbicides or biological control agents to provide longer-term management.

Annual applications of herbicides can be effective and can provide relatively successful season-long control of purple loosestrife stands. Control rates of > 90% can be accomplished with applications of the herbicides 2,4–D, glyphosate, triclopyr, imazapyr and imazamox. Single applications of registered herbicides generally do not provide satisfactory control of loosestrife for more than one season. Multi-season control of purple loosestrife can be achieved using imazapyr; however, the rates required for this level of control often have a negative impact on desirable vegetation, which limits its use. Herbicides used to control purple loosestrife have very different selectivity spectrums for nontarget plants. In addition, application rate affects selectivity. When selecting a herbicide for management of purple loosestrife, it is important to consider the impact of the herbicide on the many important nontarget wetland species that may be affected by overspray or exposure to high concentrations of herbicides needed to effectively control purple loosestrife (Chapter 11). In addition, readers should be aware that most states require application permits before herbicides can be used for management of purple loosestrife in wetlands or other aquatic locations.

The vast seedbank in the soil of established stands of purple loosestrife facilitates regrowth of the species after herbicides dissipate and are no longer effective. Therefore, the most effective long-term option for suppressing and controlling the growth of this invasive weed may be the use of biological control (Chapters 8 and 9). Research and evaluation of potential biological control agents for the North American purple loosestrife invasion identified a number of European insects that showed promise as biocontrol agents. The USDA-APHIS has now approved five European insect species for introduction as classical biocontrol agents. These include two leaf-feeding beetles [*Galerucella calmariensis* L. and *G. pusilla* Duftschmidt (Coleoptera: Chrysomelidae)], a root-mining weevil [*Hylobius transversovittatus* Goeze (Coleoptera: Curculionidae)] and a flower-feeding weevil [*Nanophyes marmoratus* Goeze (Coleoptera: Curculionidae)]. The fifth insect approved was the seed-feeding weevil *Nanophyes brevis* Boheman (Coleoptera: Curculionidae), but this insect was ultimately not introduced due to problems obtaining healthy, parasite-free insects from Europe. Initial releases of the leaf-feeding beetles *Galerucella* spp. and the root-mining weevil *Hylobius* sp. into natural areas from New York to Oregon were experimental and early observations suggested that the leaf-feeding beetles occasionally feed on native plant species; however, this now appears to be of little consequence.

G. calmariensis and *G. pusilla* are leaf-feeding beetles easily confused with native North American *Galerucella* species. The European species, however, seriously affect purple loosestrife growth and seed production by feeding on the leaves and new shoot growth. The two introduced beetles are similar in appearance and share similar life history characteristics. Adults overwinter in leaf litter and emerge in the spring shortly after shoot growth begins. Peak dispersal of overwintered beetles occurs during the first few weeks of spring, when new-generation beetles make dispersal flights shortly after emergence and can locate host patches greater than a half mile away within only a few days. Adults feed on shoot tips and females lay 2 to 10 eggs on the leaves and stems of purple loosestrife from May to July. Young larvae feed on developing leaf buds, while older larvae feed on all aboveground plant parts. Pupation by mature larvae takes place in the litter below the plant. Reports from several locations describe complete defoliation of large multi-acre stands of purple loosestrife, with local biomass reductions of greater than 95%. These results are limited and localized, but have occurred in states ranging from Connecticut to Minnesota and into the provinces of Canada to date.

Larvae of the introduced root-boring weevil *H. transversovittatus* hatch and feed on root tissue for one to two years depending on environmental conditions. Pupation occurs in the upper part of the root, with adults emerging between June and October. Adults then feed on foliage and stem tissue and can live for several years. The root-boring weevil can survive in all potential purple loosestrife habitats, except for permanently flooded sites. Adults and larvae can survive extended submergence, depending on the temperature, but excessive flooding prevents access to plants by adults and eventually kills developing larvae. Feeding by adults has little effect on the plants, but as is typical, feeding by larvae can be very destructive to the rootstock.

The flower-eating weevil *N. marmoratus* has been introduced to several states and is widespread in Europe and Asia, where it is able to tolerate a wide range of environmental conditions. The flower-eating weevil severely reduces seed production of purple loosestrife as larvae consume the flower and mature larvae form a pupation chamber at the bottom of the bud. Damaged buds do not flower and are later aborted, thus reducing purple loosestrife seed output. New-generation beetles appear mainly in August and feed on the remaining green leaves of purple loosestrife. Adults overwinter in leaf litter; development from egg to adult takes about 1 month and there is one generation per year.

Summary

The introduction of purple loosestrife into North America occurred in the early 1800s with the importation of wool containing seeds, as a favorite herb in flower gardens and from released ship ballast water. Unfortunately, this attractive plant has become one of North America's most widely dispersed and dominant nonnatives in habitats ranging from dry soils to inundated marsh areas or lakes. Stems can grow as tall as 10 feet and can form densities of up to 50 stems per plant, creating a canopy that limits light and space to native plants. Purple loosestrife causes problems in wetland ecosystems by forming dense monocultures, outcompeting native plants, altering hydrology and changing water chemistry, which all in turn affect native plant and animal communities. Purple loosestrife is an easily identified emergent plant, which facilitates hand removal and selective herbicide applications. These methods can provide temporary control of small populations, but access to the species is often limited. Populations are most effectively controlled when multiple control methods are used in conjunction, but biocontrol seems to provide the best long-term suppression of dense stands of purple loosestrife. Fortunately, classical biocontrol agents appear to be able to successfully reduce populations of purple loosestrife throughout North America.

For more information:

- Brown CJ, B Blossey, JC Maerz and SJ Joule. 2006. Invasive plant and experimental venue affect tadpole performance. Biological Invasions 8:327-338.
- Invasive plants of the eastern United States website. http://www.invasive.org/eastern/biocontrol/11PurpleLoosestrife.html
- Invasive species: purple loosestrife (*Lythrum salicaria*). Wisconsin Department of Natural Resources website. http://dnr.wi.gov/invasiveS/fact/loosestrife.htm
- Muenscher, WC. 1967. Aquatic plants of the United States. Cornell University Press.
- Purple loosestrife (*Lythrum salicaria*) in the Chesapeake Bay watershed: a regional management plan. 2004. http://www.anstaskforce.gov/Species%20plans/doc-Purple_Loosestrifel_Mgt_Plan_5-04.pdf
- Purple loosestrife: what you should know, what you can do. Minnesota Sea Grant Program (aquatic species) website. http://www.seagrant.umn.edu/ais/purpleloosestrife_info

Photo and illustration credits

Page 119: Purple loosestrife; Bernd Blossey

Page 120: Line drawing; adapted from Muenscher (1967)

Page 122: Mating pair of the leaf-feeding beetle *Galerucella calmariensis;* Bernd Blossey

Page 123: Adult root-boring weevil *Hylobius transversovittatus*; Bernd Blossey

Thomas Woolf
Idaho State Department of Agriculture, Boise ID
twoolf@agri.idaho.gov

Chapter 13.7: Curlyleaf Pondweed

Potamogeton crispus L.; submersed aquatic plant in the Potamogetonaceae (pondweed) family
Derived from *potamos* (Greek: river), *geiton* (Greek: neighbor) and *crispus* (Latin: curly)
"curly-leafed plant close to the river"

Introduced from Europe in the mid 1800s
Present in all lower 48 states; particularly problematic in northern states and Canada

Introduction and spread

Native to Europe, Asia, Africa and Australia, the first known collection of curlyleaf pondweed in North America occurred in Philadelphia in 1841. The plant spread to the Great Lakes region in the early 1900s and today is found in all of the contiguous 48 states. The spread of curlyleaf pondweed throughout the US can be attributed to boat and fish hatchery activity. Curlyleaf pondweed is now thoroughly naturalized in the United States and Canada and is considered an exotic weedy species throughout its range.

Description of the species

Curlyleaf pondweed is a rooted submersed herbaceous perennial monocot that grows in lake and river systems and aggressively outcompetes native submersed vegetation. The species has wavy leaves with finely serrated or toothed margins and a "crisp" leaf texture. Leaves are typically green early in the season and can become red when they near the water's surface. The oblong-shaped leaves are 1 to 3" in length and are attached to the stem in an alternate arrangement. Long spaghetti-like stems form as the plant quickly grows to the water's surface and develops into dense weedy mats.

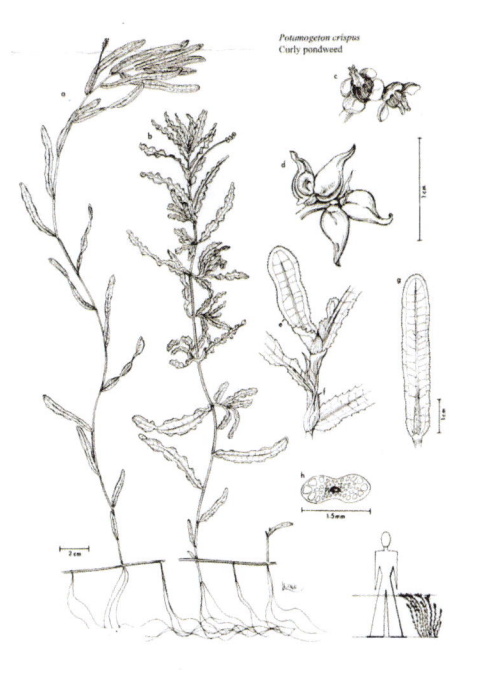

Curlyleaf pondweed grows in conditions ranging from ice-covered waters with very low light intensities to summer conditions with very warm temperatures and intense sunlight. Colonization by curlyleaf pondweed is limited by light availability and the species typically inhabits waters that range from 3 to 6 feet in depth, but curlyleaf pondweed has been found at depths of more than 20 feet in very clear water. This species prefers to grow in still water, but curlyleaf pondweed is quite tolerant of flow and is found in many river systems throughout the US and Canada.

Curlyleaf pondweed is often found in nutrient-rich or eutrophic systems and the species has a high tolerance for nutrient pollution and low light conditions. In fact, the species is sometimes considered an indicator of pollution and eutrophication due to its tolerance of low light and high dissolved nutrients.

Reproduction

Curlyleaf pondweed reproduces primarily by producing turions and rhizomes. Turions are hardened modified reproductive buds that form from apical buds, in leaf axils or directly from rhizomes prior to plant senescence in early summer. A single plant produces an average of 5 turions, with each turion averaging 4 buds. Turions constitute over 40% of the total plant biomass prior to senescence and turion densities of more than 1,000 per square foot have been reported in lake sediments. Each turion can remain viable in the sediment for multiple seasons and can sprout multiple times. Flowering usually coincides with turion formation. Flowers are very small, inconspicuous and borne on small spikes that emerge above the water surface. Seeds are produced but germination rates are quite low (0.5%). As a result, reproduction of curlyleaf pondweed is due mainly to the production and sprouting of vegetative turions.

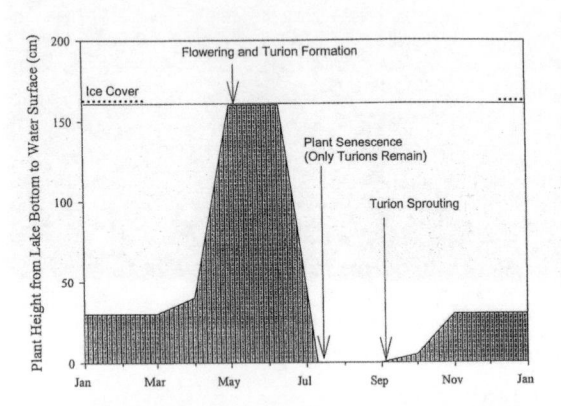

Curlyleaf pondweed has a life cycle that is fairly unique for submersed aquatic plants. Plants flower and produce turions, then die back or senesce, typically in early summer. Turions lie dormant throughout the summer and then sprout in the fall when water temperatures drop to below 66 °F and daylength shortens to fewer than 11 hours of daylight. Plants grow and can reach from an inch to several feet in height until water temperatures fall below 50 °F. When temperatures drop below 50 °F, growth of curlyleaf pondweed slows or stops and plants overwinter in a very slow-growing or dormant state. Since the species overwinters with green growth above the sediment, curlyleaf pondweed often has an advantage over native species when growth resumes in the spring.

Curlyleaf pondweed can grow up to 4" per day when days become longer and water temperatures start to rise in early spring. Plants quickly grow to the surface and turion production and flowering begin. Dense mats of curlyleaf pondweed also form on the water surface and shade out competing species. Turion production and flowering are followed by senescence or dieback, which occurs by the 4th of July in many areas.

Problems associated with curlyleaf pondweed

Curlyleaf pondweed forms dense mats on the water's surface in May and June, which inhibits fishing, boating and other types of water recreation. Dense growth of curlyleaf pondweed in moving water systems can obstruct flow and can exacerbate flooding due to large amounts of biomass obstructing river channels. Dense surface mats of plant material also limit light to low-growing submersed native species; in fact, monocultures of curlyleaf pondweed often result from this competition for light. Dense vegetation at the water's surface also can stagnate the water column and inhibit oxygen exchange from the surface to the lake bottom. Decomposing plant material under the weedy canopy further reduces dissolved oxygen levels in the water column. These conditions can reduce or eliminate fish and aquatic invertebrates in dense beds of curlyleaf pondweed. Mosquitoes, on the other hand, find curlyleaf pondweed beds to be the ideal habitat.

Curlyleaf pondweed typically senesces when water temperatures rise and dissolved oxygen levels begin to decline. The large amount of decomposing biomass produced from senescence releases nutrients and decreases oxygen in the water column, which further stresses the aquatic community. Algal blooms commonly occur after senescence of curlyleaf pondweed and decreased water clarity and oxygen levels can persist for the entire summer season.

Management options

Curlyleaf pondweed often requires management in order to preserve the recreational and environmental value of the bodies of water infested by the species. The most effective and efficient way to protect waterbodies from curlyleaf pondweed and other invasive aquatic species is prevention. Curlyleaf pondweed is on a number of state noxious weed lists, which make it illegal to sell or transport the species. The best way to prevent the introduction of curlyleaf pondweed into new waterbodies is to ensure that all plant material is removed from boats and trailers. Boats, trailers and gear should be thoroughly inspected, washed (with hot water) and dried before moving to a different water body to prevent the spread of curlyleaf pondweed and other invasive aquatic species.

There are a number of options for control and management in bodies of water that are already infested with curlyleaf pondweed. Physical (Chapter 6) or mechanical (Chapter 7) control options include hand removal, benthic barriers and mechanical harvesting. Hand removal by raking or hand pulling using divers can be effective tools for controlling plants in localized areas, but these efforts can be costly and time-intensive. The turion bank in the sediment should also be considered with hand removal, since regrowth from turions can quickly reinfest cleared areas. Curlyleaf pondweed can also be spread by fragments, so measures should be taken to prevent fragments and turions from spreading. Benthic barriers are effective for curlyleaf pondweed control in localized areas. The barriers prevent regrowth from turions in the sediment and, if barriers are maintained, can provide long-term control. However, benthic barriers are labor-intensive to install and maintain and often require installation permits. Mechanical harvesting can provide temporary control of curlyleaf pondweed, but can also exacerbate the spread of fragments and turions. Management programs can include mechanical harvesting to improve boater and recreation access by effectively "mowing the lawn" to remove nuisance growth, but disposal of harvested biomass can be problematic due to the large volumes of heavy plant material. Drawdown of a body of water is an effective method for seasonal control of curlyleaf pondweed. However, drawn-down areas of shoreline can quickly be reinfested by curlyleaf pondweed plants in deeper water and by sprouting of turions in the sediment. Also, drawdowns are non-specific and will likely damage populations of desirable native submersed plants as well.

There are currently no known insect or pathogen biocontrol agents that attack curlyleaf pondweed, but sterile triploid grass carp (Chapter 10) can provide control of the species. However, grass carp are non-specific herbivores that will eat many native plant species. Grass carp are also illegal in many states and can typically be used only in closed systems.

Several aquatic herbicides – including diquat, endothall, fluridone, penoxsulam and imazamox – can be used to effectively control curlyleaf pondweed. Diquat and endothall are contact herbicides and are relatively fast-acting, whereas the other herbicides are systemic products that are often used as whole-lake treatments and require longer contact times for control (Chapter 11).

Research has shown that early season treatments with herbicides can very effectively control curlyleaf pondweed and prevent turion production. Most native plant species are still dormant early in the spring, so treatment at this time prevents damage to many desirable native plants while providing selective control of curlyleaf pondweed. Since effective control early in the season prevents turion production, regrowth of curlyleaf pondweed is reduced the following year.

Summary
Curlyleaf pondweed is a problematic invasive submersed aquatic weed in the northern US and in Canada. The species grows and reproduces at very high rates and can quickly cover the entire surface of a body of water with dense monocultural growth. Dense growth of curlyleaf pondweed impedes recreation, reduces populations of native submersed plant species and alters the ecosystem so that it is inhospitable to fish and other fauna. Active management is often required to maintain the environmental and recreational value of water bodies infested with curlyleaf pondweed.

For more information:
•Minnesota Department of Natural Resources.
http://www.dnr.state.mn.us/aquatic_plants/submerged_plants/curlyleaf_pondweed.html
•Netherland MD, JD Skogerboe, CS Owens and JD Madsen. 2000. Influence of water temperature on efficacy of diquat and endothall versus curlyleaf pondweed. Journal of Aquatic Plant Management. 38:25-32. http://www.apms.org/japm/vol38/v38p25.pdf
•University of Florida Center for Aquatic and Invasive Plants. http://plants.ifas.ufl.edu/node/338

Photo and illustration credits
Page 125 upper: Curlyleaf pondweed infestation; Thomas Woolf, Idaho State Department of Agriculture
Page 125 lower: Line drawing; University of Florida Center for Aquatic and Invasive Plants
Page 126: Graph; Thomas Woolf, Idaho State Department of Agriculture
Page 127: Curlyleaf pondweed; Thomas Woolf, Idaho State Department of Agriculture

Toni Pennington
Tetra Tech, Inc., Portland OR
toni.pennington@tetratech.com

Chapter 13.8: Egeria

Egeria densa Planch.; submersed plant in the Hydrocharitaceae (frog's-bit) family
Derived from *Egeria* (Greek: water nymph) and *densa* (Latin: dense)
"densely growing water plant"

Introduced from South America to the northeastern US in the 1890s
Present throughout most of the US except Arizona and the upper Midwestern states

Introduction and spread

Egeria (*Egeria densa*), sometimes inappropriately referred to as *Elodea densa*, is easily confused with nonnative hydrilla (Chapter 13.1) and native *Elodea canadensis*. Physical similarities among the three species are responsible for the confusion in proper identification and, by extension, inconsistent naming. The popularity of egeria in home aquariums and ponds and its frequent use in biology classrooms are likely responsible for the widespread distribution of egeria across the US and elsewhere. Egeria has many common names (including anacharis and Brazilian elodea) and is commonly referred to as "oxygen weed" on many internet sites, where the species is touted for its ease of growth and ability to increase dissolved oxygen in freshwater aquariums and ponds. Many aquarists fail to consider the downsides of the plant's rapid growth rate and its effect on early-morning dissolved oxygen levels. Plants release oxygen during the day; however, plants respire (take up oxygen) at night and cause the lowest oxygen levels to occur in the early morning. Fish kills can occur if plant density is high enough and dissolved oxygen levels become depleted overnight due to plant respiration. Like many aquatic weeds, egeria was most likely brought to the US through the aquarium trade and the species was probably first introduced to natural waterways as a result of aquarium dumping and flooding of ornamental ponds. Some states now list *Egeria densa* as a noxious weed, which may slow commercial sales and introduction to new waterbodies. The current spread of egeria is due primarily to recreational activities such as boating, fishing and the use of personal watercraft. Similar to hydrilla, initial infestations of egeria are often found near public boat ramps, providing further evidence for this means of spread.

Description of the species

Egeria is a rooted submersed monocot that grows in a variety of fresh water bodies, including flowing and standing water. Growth of egeria is limited when the species is exposed to extremely warm (above around 90 °F) or cold (below around 40 °F) water for several weeks; however, egeria can withstand low light and low temperatures similar to Eurasian watermilfoil (Chapter 13.2). The species' limited tolerance for high water temperatures may explain the shift in species dominance from egeria to hydrilla during the summer in some Florida water bodies. Egeria has stems that are highly branched and can reach lengths of 25 feet or more due to the species' tolerance of very low light levels. The long stems from a single rooted plant commonly form a canopy near the water surface that can cover an area of six feet or more, a growth habit that is observed in other canopy-forming submersed weeds. Leaves of egeria are thin, small (1-1/2" long and 1/8" wide), lance-shaped and have minute teeth along the edges that may be difficult to see without a magnifying glass. Leaves are arranged in whorls around the stem, with each composed of four to six leaves per whorl. Leaf nodes are so densely spaced at the growing tip of the plant that they are indistinguishable, but nodes are more widely spaced near the main stem and on stems lower in the water column. Branches are borne from distinct and rather predictable locations along the stems of egeria. The number of leaves per whorl doubles or even triples (up to 12 leaves per whorl) every 8 to 12 leaf nodes, which has led some to refer to these unique regions as "double nodes." These double nodes are the only location where branches and flowers are borne along the stems.

Reproduction

Egeria is dioecious, meaning that plants bear only staminate (male) or pistillate (female) flowers. "Female" plants (with pistillate flowers) are not known to occur outside South America. In rare cases these plants are found, but sexual reproduction and seed set are extremely rare. This has resulted in widespread distribution in the US of "male" plants (with staminate flowers) which likely have little genetic variation. Egeria spreads exclusively from vegetative propagules including stems, branches and root crowns. Branches, roots, flowers and root crowns are formed along plant stems adjacent to double leaf nodes every 8 to 12 leaf whorls. Unlike several other invasive submersed plants, egeria does not produce tubers, turions or rhizomes to facilitate spread or to provide energy storage for overwintering. Instead, egeria relies on stems and root crowns for colonization and survival during inclement conditions. Closely spaced double nodes in stem tips result in the greatest potential for growth in this region, which can make management of the species difficult. Egeria can produce a new plant from each double node along a stem fragment; this, coupled with its rapid growth rate (easily growing up to 1/2" per day), allows for the rapid expansion and competitive ability of the species.

Problems associated with egeria

Egeria roots in the sediment at the bottom of the water body and grows completely underwater but forms a dense mat just under the water surface. The result is a thick canopy of vegetation that spreads over large areas and impacts recreation, property values, water quality and ecosystem function.

Dense growth of egeria entangles boat propellers and impedes navigation, which often results in the unintended spread of the species when stem fragments are created after a close encounter with a boat prop. Fragments can float for days or weeks before sinking into the sediment or being stranded along shorelines. These fragments quickly form roots, which results in new colonizations or substantial increases in plant bed size that would not occur naturally. Because egeria is largely transported by human activities, infestations tend to occur near boat launches, adjacent swimming areas, marinas and boat docks. Thick mats of surface vegetation in these areas are extremely unsightly and even dangerous for users of these facilities.

Water quality may be compromised by thick surface growth of egeria. Dense growth reduces the natural mixing of water by wind and causes an increase in surface water temperature during the summer, which is harmful to fish and invertebrates. Thick mats also provide a protected growth platform for filamentous algae that are unsightly, cause odors upon decay and can spawn large mosquito populations. Reduced wind mixing also restricts the entry of atmospheric gases (i.e., oxygen and carbon dioxide) to the water. Oxygen is necessary for fish and invertebrates, while carbon dioxide is necessary for growth of submersed plants, including algae. As with hydrilla, dense growth of egeria also causes wide daily fluctuations in pH and other water quality parameters, which makes infested waterways inhospitable to many aquatic animals.

Management options

Egeria has been sold as an aquarium plant in the US for as many as 50 years, but it has not spread through the country as quickly as other noxious species such as hydrilla. The first lines of defense to reduce the impacts of egeria are to prevent the introduction of the species to new water bodies and to limit its spread in waters that are already infested. The most efficient and effective preventative measure is to thoroughly remove plant fragments from boat trailers and watercraft before leaving an infested waterbody. In fact, removing all aquatic vegetation reduces the likelihood of spreading other nonnative species such as zebra mussels and other inconspicuous species. The cost of prevention (e.g., through signage, boat inspections, boat washing stations, etc.) is orders of magnitude less than the cost of managing existing populations because once egeria is established it is extremely difficult, and most would argue impossible, to eradicate.

Physical (Chapter 6) and mechanical (Chapter 7) controls for egeria are similar to those for other submersed weeds, largely due to their ability to establish new colonies from stem fragments. As a result, the benefits and drawbacks of various control methods are similar among the species. Hand removal and the use of benthic barriers can be selective; however, these methods are very laborious and time-intensive. Because egeria does not produce tubers or turions, the likelihood of reinfestation after benthic barriers are removed or when hand pulling is completed is reduced, provided both methods are employed with vigilance. Mechanical harvesters can clear large areas for boat navigation; however, harvesters can produce thousands of fragments that can expand the population. Since harvesters essentially mow the upper portions of the plant, the need to remove stem tips after mechanical harvesting cannot be understated; otherwise, stem tips float away and spread the plant to new habitats within a water body. In addition, multiple harvests are usually required during the peak growing season due to the rapid growth rate of egeria.

Water level drawdowns may be used where feasible to control egeria in regulated water bodies (e.g., irrigation canals and reservoirs for power generation or flood control). Duke Power Company has used drawdowns for many years to control egeria in power station reservoirs in the Carolinas and Virginia. Egeria may be the submersed aquatic weed most susceptible to drawdown and desiccation because seeds, tubers or turions are not produced to allow for re-growth; as a result, drawdown can provide control for 2 to 3 years. Plants are particularly vulnerable during winter drawdowns when dry and freezing conditions are present. The required duration of dewatering depends on various climatic and sediment conditions such as relative humidity, temperature and sediment density (the ability of soil to retain water). Disadvantages to drawdown include lack of specificity (nontarget native plants and wildlife are impacted) and loss of the water for other purposes such as hydropower, irrigation and recreation.

Although research is currently underway to identify effective and safe biocontrol agents, the only biocontrol agent currently available for reducing egeria biomass is the sterile grass carp (Chapter 10). Grass carp have been stocked following drawdown in some locations, which has led to long-term control. Sterile grass carp effectively control egeria in areas where low water temperature does not limit their feeding; unfortunately, egeria is capable of positive and sustained growth in climates cooler than those required for active grass carp feeding, so effectiveness may be limited under those conditions.

Herbicides commonly used to control egeria include the systemic herbicide fluridone and the contact herbicides copper and diquat (Chapter 11). The list of herbicides that can be used to effectively control egeria is very limited compared to those used to control Eurasian watermilfoil. Egeria is a monocot and is therefore not susceptible to 2,4–D or triclopyr. Endothall effectively controls hydrilla, a species that is closely related to egeria; however, endothall has no effect on egeria. Egeria is often found in systems with flowing water, which makes the use of slow-acting systemic herbicides challenging because plants require a long exposure time in order for systemic herbicides to provide effective control. The growth of egeria in flowing water systems coupled with a limited number of effective herbicides make egeria a difficult plant to control with herbicides.

Summary

The popularity of egeria in the aquarium trade and in biology classrooms has substantially contributed to its widespread distribution in the US, Europe, Asia, New Zealand, Japan, Chile, Mexico, Canada and Australia. The spread of egeria between water bodies is largely due to trailered boats and other watercraft that transport fragments. Long-lived stem fragments are easily spread by currents and watercraft within infested water bodies. When these fragments come into contact with sediments on the lake bottom or the margins of the water, the fragments form roots, plantlets develop and new colonies of egeria rapidly become established. Egeria tolerates a wide range of water quality characteristics, sediment nutrient levels and light levels and commonly grows in similar habitats favorable to Eurasian watermilfoil. As a result, it is likely that egeria can invade and colonize areas that currently support growth of Eurasian watermilfoil.

For more information:
•Bini LM and SM Thomas. 2005. Prediction of *Egeria najas* and *Egeria densa* occurrence in a large subtropical reservoir (Itaipu Reservoir, Brazil-Paraguay). Aquatic Botany 83:227-238.
•California Department of Boating and Waterways. 2000. Draft environmental impact report for the *E. densa* control program. Vol. II: Research Trial Reports.
•Cook CDK and K Urmi-König. 1984. A revision of the genus *Egeria* (Hydrocharitaceae). Aquatic Botany 19:73-96.
•Getsinger KD and CR Dillon. 1984. Quiescence, growth and senescence of *E. densa* in Lake Marion. Aquatic Botany 20:329-338.
•Pennington TG. 2007. Seasonal changes in allocation, growth, and photosynthetic responses of the submersed macrophyte *Egeria densa* Planch. (Hydrocharitaceae) from Oregon and California. PhD dissertation. Portland, OR: Environmental Sciences and Resources, Portland State University.
•University of California at Davis, Agriculture and Natural Resources. 2009. http://ucce.ucdavis.edu/datastore/detailreport.cfm?usernumber=43&surveynumber=182
•USDA NRCS. 2009. The PLANTS Database. National Plant Data Center, Baton Rouge, LA. http://www.plants.usda.gov/

Photo and illustration credits
Page 129: Egeria infestation; Toni Pennington, Tetra Tech, Inc.
Page 130: Line drawing; University of Florida Center for Aquatic and Invasive Plants
Page 131: Egeria; Toni Pennington, Tetra Tech, Inc.
Page 132: Egeria; Toni Pennington, Tetra Tech, Inc.

Chapter 13.8

Jack M. Whetstone
Clemson University, Georgetown SC
jwhtstn@clemson.edu

Chapter 13.9: Phragmites—Common Reed

Phragmites australis (Cav.) Trin. Ex Steud.; emergent plant in the Poaceae (grass) family
Derived from *phragma* (Greek: fence) and *australis* (Latin: southern)
"southern plant with fence-like growth"

Invasive variety probably introduced from Europe to the Atlantic Coast in the late 1800s (Non-invasive varieties are native)
Present throughout all states in the continental US

Introduction and spread

Phragmites (also called common reed) is a wetland species that grows from a thick, white, hollow root (rhizome) system buried deep in the substrate in areas with fresh to brackish water. The species is distributed in temperate zones throughout the world and can be found on every continent except Antarctica. Phragmites is widely distributed in North America, occurring in all US states except Alaska, and in all Canadian provinces and territories except Nunavut and Yukon. Phragmites has been widespread in the northeastern US for many years and is currently spreading west into the Great Plains. Nebraska has initiated a multi-million dollar control program on the Platte River, where growth of phragmites is totally altering the aquatic ecosystem and causing problems for endangered birds (Chapter 4). There are many distinct genotypes (varieties) of phragmites, including at least two native varieties and a nonnative variety from Europe that is much more invasive than native varieties. The European variety was probably introduced to the Atlantic Coast in the late 1800s and has expanded its range throughout North America, most notably along the Atlantic Coast and in the Great Lakes area. The European variety has replaced native plants in New England and has become established in the southeastern US, where native phragmites has historically not occurred or has been present only in small populations. European phragmites sprouts, survives and grows better in fresh and saline environments than native phragmites. The species has been called an "ecosystem engineer" because numerous changes can occur when phragmites invades an area and replaces other vegetation. Large monotypic (single-variety) stands of European phragmites are associated with decreased plant diversity. In addition, soil properties, sedimentation rates, bird and fish habitat use and food webs may be altered when marshes are converted from diverse plant communities to dense, monotypic stands of phragmites.

Phragmites is most common on wet, muddy or flooded areas around ponds, marshes, lakes, springs, irrigation ditches and other waterways. The species persists during seasonal drought as well as frequent, prolonged flooding. Phragmites tolerates brackish and saline conditions, and the invasive European variety is better adapted to areas with higher salinity than are native varieties. The species grows best in sites with fresh to low brackish water (0 to 5,000 parts per million salinity), but can reportedly survive in areas with salinities equal to full strength ocean water (35,000 parts per million). Phragmites establishes and grows well on disturbed sites and is often considered a weedy or nuisance species. The species rapidly colonizes and forms monotypic populations in disturbed areas, but is slower to colonize and dominate in diverse vegetated wetlands. Phragmites grows best in full sun and is intolerant of shade.

Description of the species
Phragmites is a robust perennial grass that may reach 20 feet tall, but generally reaches a height of 10 to 12 feet. Maximum height is usually attained when plants are 5 to 8 years old. Phragmites spreads primarily by vegetative means via stolons and rhizomes and produces dense monotypic stands of clones, or plants that are genetically identical to one another. Clones are long-lived and can reportedly persist for over 1,000 years. Phragmites produces stout, erect, hollow aboveground stems from rhizomes that persist when stems and leaves die back during winter. Stems are usually unbranched and bear leaves that are arranged in an alternate manner along the top half of the stem. Leaf blades are blue-green to green in color and have margins that are somewhat rough. Leaves are flat at maturity and measure 4 to 20 inches long and 0.4 to 2 inches wide.

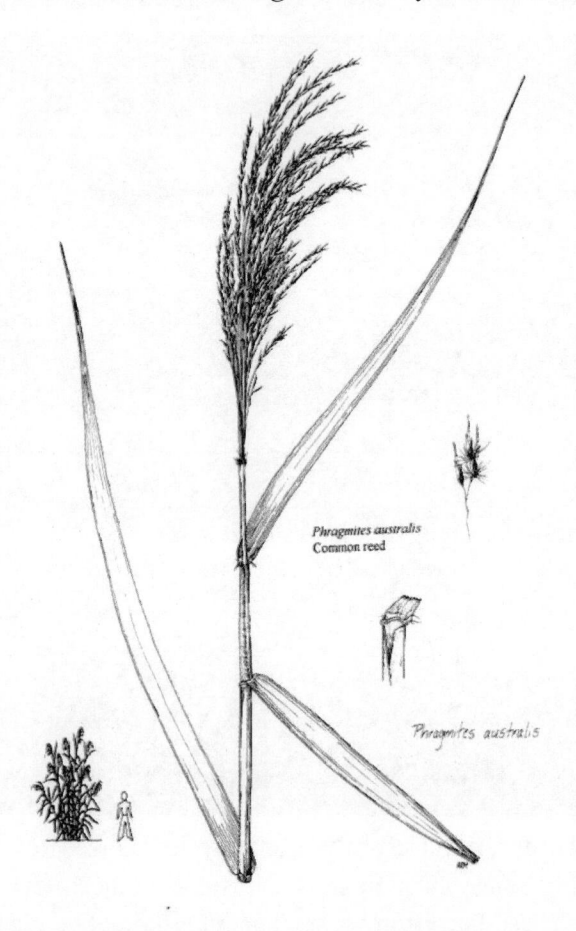

Reproduction
Phragmites reproduces sexually from seed, but most growth is from stolons (creeping aboveground stems) and rhizomes (underground stems). Stolons can grow to greater than 40 feet in length and are typically produced when water availability is low. Rhizome production and vegetative spread can be extensive and allow the species to spread into sites unsuitable for establishment from seeds. The species is often dispersed through the transport of rhizome fragments and the movement of soil or sod. Phragmites flowers are produced during mid-summer to fall and are borne in a large, feathery seed head that is 6 to 20" long. Seeds are dispersed by wind and water.

Problems associated with phragmites

Phragmites forms large monotypic stands that are virtually impenetrable. These stands replace diverse native plant communities and reduce plant, fish, bird and wildlife ecosystem productivity and diversity. However, phragmites does provide minor shade, nesting and cover habitat for mammals, waterfowl, song birds and fishes. Phragmites provides food as well as nesting, roosting and hunting habitats to a wide variety of bird species, including ducks. In addition, waterfowl, pheasants and rabbits use the margins of stands of phragmites as cover to hide from predators. Some reports suggest that immature plants are readily eaten by goats, cattle and horses, but the species is not considered a high-value or highly palatable food for livestock or wildlife when plants are mature.

Habitat use by fish, crustaceans and other aquatic invertebrates can be affected by dense growth of phragmites. For example, small fish and crustaceans prefer habitats with smooth cordgrass (a shorter and less dense native species) to those with infestations of phragmites, and populations of aquatic invertebrates are generally highest in areas with other native vegetation such as cattail. Also, several studies report that marshes dominated by phragmites provide less suitable habitat for larvae and small juvenile forms of mud minnow.

Management options

As with any invasive aquatic plant, preventing the establishment of phragmites is the best available option. This can be challenging because native and European phragmites are almost indistinguishable from one another and identification of the varieties of phragmites can only be done by experts. The range of the invasive European variety of phragmites appears to have been expanded by the movement of equipment used in ditching, drainage and dredging operations. Inspection and cleaning of equipment should be part of the operator's general protocol before moving equipment into new areas to prevent the dispersal of any aquatic invasive plants, but particularly invasive varieties of phragmites.

The use of chemical, mechanical, physical and integrated control methods are acceptable tools for the control of phragmites. There are native populations of phragmites in some areas and managers may wish to go to the expense of determining whether their populations are native plants or the invasive European variety before treating the area. Positive identification of the invasive variety requires the use of genetic tools and DNA analyses, which are currently not readily available to the public. It may be desirable to maintain and encourage populations of native phragmites while discouraging populations of the invasive European variety. For example, phragmites can be useful for erosion prevention and bank stabilization and can actually increase the elevation of some areas by trapping sediments and building "land" from decomposed plant material and root mat formation each winter. Integrated management that employs multiple control methods may lead to the most efficient and economical control plan. Mechanical (Chapter 7) and physical (Chapter 6) controls (primarily mowing and burning) have been utilized for many years, but have provided varying degrees of success and usually result in temporary control at best. There are no biological control options available to control phragmites, although large herbivores such as goats have been used to control phragmites along the Platte River in Nebraska. In addition, herbicide control options are few and only recently have new herbicides that provide medium- to long-term control been identified and registered.

Because phragmites is an emergent plant that does not grow in deep water, some control has been noted in areas that are dredged to deepen the body of water to a minimum of five to six feet. This deepening removes plants and their rhizome systems and offers long term control. However, deepening is very expensive and eliminates desirable native plants as well. In addition, the permitting required to employ this control measure is tedious and difficult.

Burning – either alone or in combination with deep flooding or herbicides – has provided some level of success in some areas. Burning alone offers only a short-term solution, especially in wet areas, because this method does not effectively control the rhizome system and can actually stimulate rhizome growth that benefits from nutrients released during burning. A multi-stage process of burning followed by deep flooding or herbicide application after plants begin to regrow has been more successful. However, parameters such as the optimum depth of flooding required and the best stage of plant growth before herbicides can be applied are unclear. Also, the use of fire to control phragmites has become impractical in many locations and permits are sometimes difficult to obtain.

Managers of some impounded areas have flooded impoundments with high-salinity water and maintained flood conditions for an extended period of time. Partial control has been obtained using this method, but a minimum of half-strength seawater (18,000 parts per million) or higher is required. The use of high-salinity flooding is extremely site-specific. Also, the invasive European variety of phragmites is more tolerant of high salinity than are native phragmites.

No purposeful introductions of insects, pathogens or diseases have been attempted to control European or native phragmites to date. Several nonnative insect species have been accidentally brought into the country with European phragmites when it was used as packing material in shipments, but these do not appear to be viable biocontrol candidates. Livestock grazing (e.g., goats, cattle and horses) on young plants of phragmites reportedly provides some control of the species. However, the nutritional value of phragmites is only fair and the logistical and health aspects associated with managing livestock in marshy, wetland situations is extremely site-specific and generally impractical.

Herbicides currently labeled for control of phragmites in aquatic habitats are the systemic herbicides glyphosate, imazamox, imazapyr and triclopyr. Glyphosate and imazapyr are broad-spectrum herbicides that control both grasses and broadleaf plants, whereas imazamox and triclopyr are selective and cause damage only to certain groups of plants. The criteria for herbicide selection are site-specific and dependent on environmental conditions, growth stage of the plant, presence of desirable nontarget plant species in the area and alternate uses of the water such as drinking and irrigation (Chapter 11).

Several general application recommendations apply for any herbicide selected. The area to be treated should be drained if possible to allow the herbicide to contact as much of the plant as possible. Also, the maximum volume of water recommended on the label should be used for herbicide applications to ensure complete coverage of all leaves and stems. Deeply flooded areas should be treated at the highest herbicide rates allowed on the label. Because phragmites occurs in large, poorly accessible, expansive areas, aerial applications may offer the most efficient and economical method of application. Additional aerial application restrictions according to the specific herbicide labels must be followed.

Backpack sprayers can be used for small infestations and spot treatments. Plants should be carefully sprayed to wet, but runoff should be avoided. Herbicide labels list more specific instructions on herbicide mixing and use.

Summary

Phragmites is a widely distributed wetland species with both non-invasive native varieties and an invasive European variety in the US. The European variety has replaced native plants in New England and has become established in the southeastern US, where native phragmites has historically not occurred. The European variety of phragmites is more competitive than native varieties and sprouts, survives and grows better in fresh and saline environments than native phragmites. The invasive nature of European phragmites results in large monotypic populations of the species, which are associated with decreased plant diversity and changes to the ecosystem that include alterations of soil properties, sedimentation rates, bird and fish habitat use and food webs. A variety of methods can be used to provide varying levels of control of invasive phragmites and the greatest success is realized when a number of different methods are employed in an integrated program. However, control of the invasive European variety of phragmites is made more challenging by the presence of the native non-invasive varieties, which can be a desirable part of aquatic ecosystems.

For more information:

- Common reed management. Texas A & M University. Texas Agrilife Extension Service.
http://aquaplant.tamu.edu/database/emergent_plants/common_reed_mgmt.htm
- Common reed: *Phragmites australis*. University of Florida Center for Aquatic and Invasive Plants.
http://plants.ifas.ufl.edu/node/323
- Environmental assessment for control of *Phragmites australis* in South Carolina. US Army Corps of Engineers.
http://www.sac.usace.army.mil/assets/pdf/environmental/Final_Phragmites_EA.pdf
- *Phragmites australis*. United States Department of Agriculture – US Forest Service.
http://www.fs.fed.us/database/feis/plants/graminoid/phraus/all.html

Photo and illustration credits:

Page 135: Common reed; Ann Murray, University of Florida Center for Aquatic and Invasive Plants
Page 136: Line drawing; University of Florida Center for Aquatic and Invasive Plants
Page 138: Common reed; Ann Murray, University of Florida Center for Aquatic and Invasive Plants

Marc D. Bellaud
Aquatic Control Technology, Inc., Sutton, MA
mbellaud@aquaticcontroltech.com

Chapter 13.10: Flowering Rush

Butomus umbellatus L; emergent shoreline plant in its own family, Butomaceae (flowering rush); originally placed in the *Alismaceae* (water-plantain) family

Derived from *bous* (Greek: ox) and *temno* (Greek: "I cut"), referring to its sword-like leaves with sharp edges that cut the mouths of cattle feeding on the species

First identified along the St. Lawrence River in Quebec in 1897; likely introduced from Europe as a garden plant
Present in the northern US from Idaho to Maine and in the adjacent Canadian provinces

Introduction and spread

Flowering rush (*Butomus umbellatus* L.) is native to Europe and Asia. It is thought that the species was first introduced to the US for use in ornamental gardens, but flowering rush thrives along shallow shorelines and in wetlands. The first observation of the species in North America occurred along the St. Lawrence River in Quebec in 1897 and botanists believe that multiple introductions have occurred since that time. By the mid 1950s, flowering rush populations were documented throughout the Great Lakes Region. Populations of flowering rush in the Great Lakes and points west are believed to be of European origin, whereas populations in the St. Lawrence River area are thought to be from Asia. Since the 1950s, flowering rush has spread to the west, north and east of the Great Lakes, with populations now found across the northern US and extending from Washington to Maine and nearly all of the adjacent Canadian provinces. Flowering rush tolerates a wide variety of shallow water and wetland settings and often forms dense stands that displace native riparian species, degrade fish and wildlife habitat, alter hydrologic patterns and interfere with recreational use of water bodies.

Description of the species

Flowering rush is a perennial monocot herb that can reach up to 5 feet in height and tolerates a wide variety of riparian and wetland habitats. Plants have an extensive rhizome and root system and soil type or consistency and soil pH do not appear to effect growth. However, the species cannot grow in shade and is intolerant of saline or brackish waters. Plants become established in wet areas or along the shallow margins of lakes, ponds and streams and can grow into water up to 9 feet deep. Leaves of flowering rush are fleshy, thin and sword-like and resemble those of native bulrush (*Sparganium* spp.), but are triangular in cross-section. Submersed leaves remain limp or float on the surface of the water, whereas emergent leaves can reach to 3 feet in length and may have tips that are twisted in a spiral manner. Flowering rush is easiest to identify when it is flowering, which only occurs if plants are growing in very shallow water or along the shoreline. Plants flower between June and August, depending on temperature and latitude. The flowers are borne in an umbrella-shaped cluster (umbel). Individual flowers have three petals that are white to pink to purple in color.

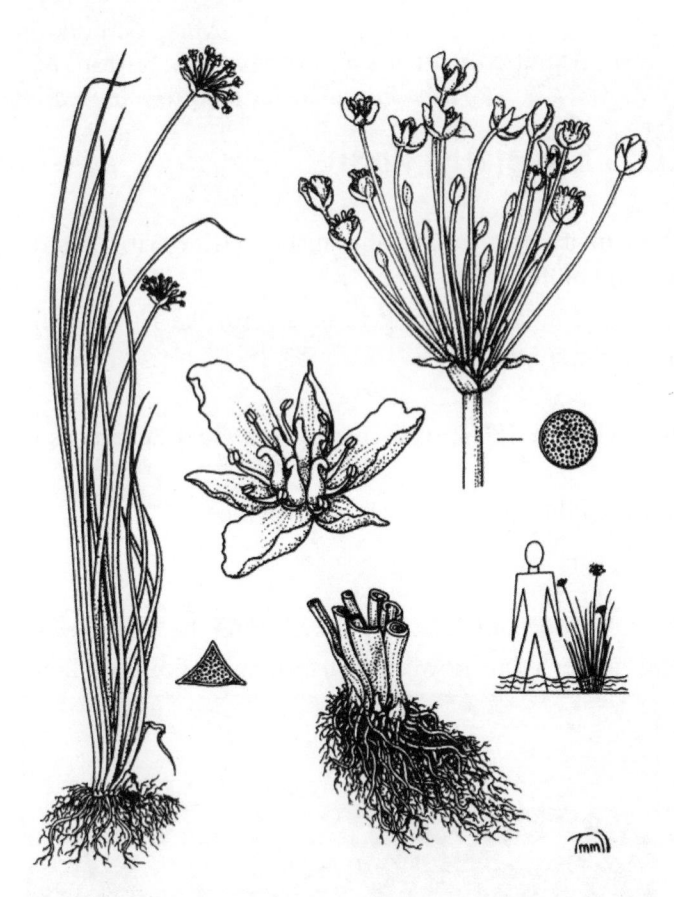

Reproduction

Flowering rush is dispersed in four ways: seeds, vegetative bulblets produced on the inflorescence at the base of flower stalks, vegetative bulblets that form along the sides of rhizomes (underground stems with nodes that produce new shoots and roots), and rhizome fragments. Once established, the species expands its population size and spreads locally by rhizome elongation. Both seeds and bulblets can be transported by water currents and are long-lived, which facilitates their dispersal by wildlife, boaters and other human activities.

Eastern US populations of flowering rush are reportedly fertile diploids (with 2 sets of chromosomes), whereas sterile triploid populations (with 3 sets of chromosomes) occur in western North America. Diploid populations flower prolifically and produce both seeds and bulblets and their spread is due to dispersal of seeds and bulblets. Triploid populations in the West rarely flower and produce low numbers of seeds and bulblets. As a result, the majority of the spread of western populations is due to rhizome fragmentation, which results in clonal (genetically identical) populations.

Problems associated with flowering rush

Flowering rush can form dense infestations that compete with native riparian species and displace more desirable plants. Dense growth of the species may also allow it to outcompete threatened or endangered plant species and likely alters wildlife habitats. There are varying levels of concern about the impact of flowering rush on wetlands and fresh water habitats. For example, reports from the St. Lawrence River suggest that even high densities of flowering rush have not significantly reduced plant diversity. However, displacement of native plant species and the potential for wildlife habitat alteration make flowering rush a species of concern.

The impacts of flowering rush to water use and access may be more significant. For example, flowering rush has developed extensive monotypic populations in reservoirs with widely varying water levels in western states. The species is also currently causing economic impacts in irrigation canals and drainage ditches in the western US and large populations of flowering rush impede access to shallow lakes by colonizing shoreline areas where aquatic plants have not grown in the past. Marshlands are becoming dominated by flowering rush because the species thrives in areas with fluctuating water levels and expansion throughout littoral zones interferes with shoreline access, boating and fishing.

Management options

Unlike many other invasive species, there is not a wealth of information regarding the management of flowering rush infestations in North America. However, the same management philosophies hold true – early detection of introductions and rapid response to new infestations provide the most effective control of flowering rush and limit further spread of the species. Flowering rush resembles many native species; therefore, accurate identification of the species is critical before initiating management efforts to avoid damaging nontarget desirable native plants.

Manual control methods include cutting and hand digging (Chapter 6). Cutting will not kill flowering rush because the species will produce new growth from underground roots and rhizomes, but this method may decrease abundance and prevent seed and bulblet production by removing inflorescences. Plants should be cut below the water surface and care should be taken to remove all cut plant parts from the water. Multiple cuts may be required throughout the summer to provide adequate control and to prevent the formation of flowers, seeds and bulblets. Hand digging is useful only when managing individual plants or small infestations. The entire root structure must be carefully removed because fragments of roots, rhizomes or bulblets left in the sediment can rapidly regrow. All plant parts removed during cutting or hand digging must be taken out of the water and transported well away from water or wetland areas to prevent recolonization.

The use of herbicides to control flowering rush is challenging due to the limited foliage available for herbicide coverage and uptake. Often only a small part of the plant emerges above the water and foliar herbicide coverage is so limited that herbicides are generally not very effective. The best time to apply foliar herbicides is likely during periods when water levels are low to improve herbicide coverage. There is no product that is selective for flowering rush and controls the species without the potential for harming other plants, so care must be taken during herbicide application to avoid impacts to nontarget species. Research by the Minnesota Department of Natural Resources suggests that a mid-summer treatment with imazapyr may be effective and research on management of this invasive weed is ongoing.

Summary
Flowering rush is an invasive species that has steadily expanded its range across the northern US and the Canadian provinces. It closely resembles bulrush and other native species and is difficult to identify unless it is flowering. The species employs multiple reproductive strategies that have helped to expand its range over the past 50 years. All the potential impacts of this invasive species on aquatic systems are not yet known, but flowering rush is capable of abundant growth that can displace native species and alter habitats. Also, dense shoreline growth of the species can certainly interfere with access and recreational uses of infested water bodies. There is limited information available regarding the management of flowering rush, but as with other invasive species, early detection and rapid response are paramount to successfully controlling new infestations. Cutting below the water surface, careful hand-digging and selective treatment with herbicides are currently the most effective strategies to control infestations of flowering rush. The expansion of flowering rush has occurred primarily in the western US and it is difficult to predict how extensive the problem may become, but research is underway to investigate the biology of the species and to identify management options that may be useful to control the spread of flowering rush.

For more information:
•Crow GE and CB Hellquist. 2000. Aquatic and wetland plants of northeastern North America. University of Wisconsin Press.
•Minnesota Sea Grant Aquatic Invasive Species website.
http://www.seagrant.umn.edu/ais/floweringrush
•Oregon Department of Agriculture, Plant Division, Noxious Weed Control website.
http://www.oregon.gov/ODA/PLANT/WEEDS/profile_floweringrush.shtml
•Rice P, V Dupuis and S Ball. Flowering rush: an invasive aquatic macrophyte infesting the flathead basin (PowerPoint). http://www.weedcenter.org/Newsletter/rice_floweringrush_sshow.pdf
•University of Florida Center for Aquatic and Invasive Plants
http://plants.ifas.ufl.edu/node/75
•USDA NRCS. 2009. The PLANTS Database. National Plant Data Center, Baton Rouge, LA.
http://www.plants.usda.gov/java/profile?symbol=BUUM
•US Forest Service Invasive Plants website.
http://www.na.fs.fed.us/fhp/invasive_plants/weeds/flowering-rush.pdf

Photo and illustration credits:
Page 141: Flowering rush; Thomas Woolf, Idaho State Department of Agriculture
Page 142: Line drawing; University of Florida Center for Aquatic and Invasive Plants
Page 143: Flowering rush; Thomas Woolf, Idaho State Department of Agriculture

Appendix A

Carlton Layne and Don Stubbs
US Environmental Protection Agency (retired)
Layn1111@bellsouth.net
donald271@verizon.net

Requirements for Registration of Aquatic Herbicides

History of pesticide regulation

A pesticide is defined as any product that claims to control, kill or change the behavior of a pest. The United States first started regulating pesticides in 1910. The 1910 Federal Insecticide Act was intended to protect farmers from adulterated products and false labeling claims. With the continuous increase in pesticide development and use after World War II, Congress passed the Federal Insecticide, Fungicide and Rodenticide Act (FIFRA) in 1947. This act, which would be amended through the years, required that all pesticides be registered with the Department of Agriculture before they could be shipped in interstate commerce. The same federal agency responsible for agricultural production in the United States was now responsible for the regulation of pesticides on agricultural crops. FIFRA established procedures for the registration and labeling of pesticides, but dealt mainly with the efficacy or effectiveness of pesticides and did not regulate pesticide use. Almost anyone could use a pesticide for any purpose and there was no legal recourse if a pesticide was not properly used. In addition, FIFRA did not allow for the denial of a pesticide registration request.

In 1962 Rachel Carson published "Silent Spring," which drew widespread public attention to the indiscriminate use of pesticides with unknown human health and environmental effects. Many of the pesticides were persistent in the environment and were transferred from one animal to the next upon being eaten (a phenomenon known as bioaccumulation). As a result, some pesticides were ultimately ingested by humans and other nontarget animals, including wildlife. Very little was known at the time about the fate of pesticides in the environment and the potential effects of their residues on man and wildlife.

The Environmental Protection Agency (EPA) was created in 1970 and the responsibility for regulating pesticide use and labeling was transferred from the USDA to this new agency. This marked the beginning of a shift in the focus of federal policy from the control of pesticides for reasonably safe use in agricultural production to the control of pesticides for the reduction of unreasonable risks to man and the environment. In 1972 Congress passed the Federal Environmental Pesticide Control Act, which amended FIFRA and set up the basic American system of pesticide regulation to protect applicators, consumers and the environment that we have today. This Act gave the EPA greater authority over pesticide manufacturing, distribution, shipment, registration and use. EPA could now, among other things:

1) require additional data as necessary;
2) suspend or cancel the registration of existing pesticides;
3) prohibit the use of any registered pesticide in a manner inconsistent with label instructions;
4) require that pesticides be classified for specific uses;
5) deny a registration request;
6) provide penalties (fines and jail terms) for violations of FIFRA;
7) provide states with the authority to regulate the sale or use of any federally registered pesticides in that state as long as state rules were at least as strict as federal guidelines.

In 1988 Congress once again amended FIFRA by requiring the EPA to reregister all pesticides registered before November 1984 and to ensure that the database was current and in accordance with modern science. The development of the Food Quality Protection Act (FQPA) in 1996 amended both the FIFRA and the Federal Food, Drug, and Cosmetic Act (FFDCA). This Act set a single health-based standard for residues of pesticides in food and required the EPA to reevaluate all tolerances for pesticides and their inert ingredients.

Registration

Pesticide regulations are continuously under review and revision as scientific methods and knowledge increase. The following sections of this chapter will discuss pesticide registration and enforcement of pesticide laws, which are just a portion of the EPA's overall responsibility to protect the environment. It costs 30 to $60 million or more, and 8 to 10 years, to introduce a new pesticide to the market. Pesticides that are destined for use in aquatic systems in the US must be registered by the federal government through the EPA and by the state in which the pesticide will be used. The product may only be used in accordance with the label accepted by the EPA and any other applicable state regulations as long as the state regulations are at least as restrictive as the federal label. A pesticide may occasionally be registered by a state based on a special local need. In such circumstances, the active ingredient of the pesticide must be registered by the EPA and the appropriate tolerances in fish, shellfish and irrigated crops must be established by the EPA. This federal agency has overall responsibility for pesticide regulation even in states with small but locally important pest control needs.

The burden of proof to show that a pesticide will not cause unreasonable adverse effects on man and the environment rests with the registrant (the company that develops or labels the pesticide). The registrant is responsible for testing the active ingredient and the end use product (the final formulated product offered for sale) for potential harm to man and the environment. The EPA requires between 84 and 124 different studies to satisfy this requirement. These studies include toxicity and exposure tests on laboratory animals that measure the possible effects of the pesticide on human health – to applicators and to the general public – through direct exposure and through residues in food. These studies also determine the fate of the pesticide once it is introduced into the environment and the effect of the pesticide on nontarget organisms. The EPA reviews these studies and determines the appropriate labeling for the use of each pesticide. Label precautions may include user safety information (protective clothing, reentry intervals or specific hazards), environmental safety warnings, container disposal and pesticide classification. In addition, all labels must provide appropriate directions for use (see "Pesticide Labeling" below).

The EPA regulates pesticide use from occupational (applicator/worker), residential and dietary standpoints and determines the potential effects of acute (immediate), intermediate and chronic (long term) exposure to humans. If the use of a pesticide results in a residue of the pesticide in food or feed, it is necessary to establish a tolerance level for that pesticide under the FFDCA. The EPA also evaluates residues in drinking water and must determine whether pesticide residue levels found in drinking water, fish, shellfish and any other food or feedstock meet the safety standard of the FQPA. In short, the EPA verifies that there is a reasonable certainty that no harm will result from the residues of the pesticide in food or feed. The FQPA is a risk-based statute and does not provide for the analysis of risks vs. benefits. Examples of some of the studies required before a product can be used as a pesticide are listed below. More detailed information is available at: http://www.epa.gov/lawsregs/search/40cfr.html. Click on "Chapter I"; then under "Browse Parts" click "150 – 189"; and finally under "Table of Contents" click "158.1 to 158.2300."

Toxicity studies (how dangerous is the pesticide to humans?)
- Acute toxicity: study the immediate effects of exposure to determine appropriate user precautions
- Sub-chronic toxicity: examine intermediate toxicological effects to identify the risks of less than lifetime exposure
- Chronic toxicity: evaluate long-term toxicity effects to determine possible problems associated with a lifetime of exposure
- Oncogenicity: determine whether the product causes cancer
- Developmental and reproductive toxicity: identify any effects on development and reproductive function

Chemistry studies (what is the pesticide?)
- Chemical identity, physical and chemical properties
- Disclosure of manufacturing process and all inert ingredients
- Determine chemicals of concern including the active pesticide and inert components
- Develop analytical methods for determining concentrations of the pesticide in plants, soil, water and food
- Determine the amount of pesticide left on plants, soil, water and food as a result of use

Environmental fate (what happens to the pesticide after it has been applied?)
- Hydrolysis: establish the significance of chemical breakdown in water
- Photolysis: determine the interaction of the pesticide with light
- Degradation: determine when the pesticide breaks down and what it breaks down to in water, soil and air
- Metabolism: examine the breakdown of the pesticide by organisms in the soil and water
- Mobility and bioaccumulation: determine how the pesticide moves in the environment and whether it accumulates up the food chain
- Field dissipation: test and monitor how the pesticide behaves under realistic conditions

Ecological toxicity (how dangerous is the pesticide to fish, birds, mammals and plants?)
- Acute toxicity: study the immediate effects on wildlife
- Chronic dietary toxicity: examine the effects of a lifetime of exposure in birds
- Reproduction studies
- Toxicity to plants

Because the EPA relies on data submitted by the registrant, it carries out a laboratory audit program. This program sends EPA scientists and enforcement personnel to laboratories that conduct studies on pesticides. These personnel are responsible for reviewing the testing procedures to ensure that they are carried out in accordance with EPA regulations for conducting good laboratory studies. In addition, the EPA requires the registrant to submit to them any data concerning adverse effects associated with the use or new testing of the chemical. These data are immediately reviewed by the EPA and any corrective action (label changes, use deletions or product cancellation) is taken as deemed necessary by the agency.

Tolerances

A tolerance is a residue level established by regulation which is considered a "safe level" of a pesticide and it is also an enforceable level. An "enforceable level" essentially means that when a pesticide is found in or on a food product and is either (1) not registered for use on that food product, or (2) present at a level higher than the tolerance established for that food crop, the food crop may be destroyed and investigations must be conducted to determine whether fines or other penalties are warranted. The tolerance is based on acute and chronic animal toxicity data. These data are multiplied by a 100-fold safety factor to determine an allowable residue level. The EPA does not set tolerances in drinking water as a result of pesticide use, but it does assess the safety of drinking water using the same safety standard for water as it does for food or feed before it will register the pesticide. Under the FFDCA as amended by the FQPA in 1996, a tolerance may only be established when EPA determines that there is a reasonable certainty that no harm will result from the aggregate exposure (food, water and residential exposure) to the active ingredient and the inert ingredients in the pesticide.

Pesticides that are registered for use in a way that results in residues of the pesticide or its metabolites of concern in or on food or feed require the establishment of a tolerance under the FFDCA. Tolerances for pesticides are established under the FFDCA by the EPA. Food or feed contaminated with residues of pesticides or their metabolites of concern that do not have an established tolerance or have residues above the established tolerance level are considered adulterated and may be seized and destroyed by the Food and Drug Administration (FDA). While the EPA sets these pesticide tolerances, the FDA is responsible for enforcing them. Pesticides to be used in aquatic systems must have established tolerance levels of that pesticide and its metabolites of concern in fish, shellfish and any crops that would be irrigated with treated water.

Pesticide labeling

Pesticides are classified as either "general use", which can be purchased and used by anyone, or "restricted use", which may only be sold to and used by persons under the direct supervision of a certified applicator. A certified applicator must complete the appropriate federal or state training and testing. Pesticides can be used to control nuisance aquatic weeds without causing unreasonable adverse effects to man or the environment as long as label directions, precautions and warnings are followed.

The EPA regulates pesticides through pesticide labeling and determines the appropriate minimal label information required for the safe and effective use of the pesticide based on data submitted by the registrant. All labels must also include certain information; for example, all labels must carry several specific statements including "Keep Out of Reach of Children" and a signal word (Caution, Warning or Danger). Directions for use – including application rates, number of applications allowed per season, user precautions, environmental precautions, container disposal instructions and other directions as determined by the EPA – are also required. In addition, every label must carry the statement "It is a violation of Federal law to use this product in a manner inconsistent with its labeling." This means the pesticide can only be used in accordance with the label on the product container. The EPA stamps the label as accepted and this is the only label the registrant may place on its pesticide container before selling the product to the public. This label then becomes the principal communication between the registrant and the user. The directions for use, precautions and warnings tell the user how to use the pesticide and what precautions to take when the pesticide is used. Any changes to the labeling must be submitted to and approved by the EPA prior to marketing. For a full discussion on labeling requirements, please visit the EPA website on labeling at http://www.epa.gov/oppfead1/labeling/lrm/

Review of registered pesticides

In 2008 the EPA completed its reregistration of all pesticides registered prior to November 1984 as required by the 1988 amendment to FIFRA. This effort took over 20 years as it required the reassessment of all products and their associated tolerances. In 2008 the EPA also initiated a Registration Review Program. This program, required by the 1996 amendments to FIFRA (FQPA), will review the registration of all registered pesticides on a continual 15-year cycle to ensure that pesticides remain in compliance with developing changes in science, public policy and pesticide use practices. For more information about the Registration Review Program go to http://www.epa.gov/oppsrrd1/registration_review/highlights.htm

Enforcement

To ensure compliance with the requirements of FIFRA, federal agents and state inspectors monitor the marketplace and conduct inspections and investigations at establishments where pesticides are produced and distributed and at facilities of commercial and private applicators where pesticides are stored. While all enforcement efforts are important, use-related inspections and investigations provide ongoing feedback to the EPA regarding the effectiveness of label requirements and accepted directions for use. This information, coupled with the requirement that registrants report all unanticipated adverse effects encountered as part of the distribution, sale and use of a pesticide, provides an impetus for additional data requirements. Mandatory label modifications may also be ordered depending on the nature of the data received.

It is a violation of federal law for any person to use any registered pesticide in a manner inconsistent with label directions. The directions can cover all aspects of the pesticide, including transportation, storage, mixing, loading, application rates, target pests, use sites or crops, methods of application, personal and worker protection, environmental warnings, disposal and anything else necessary to protect human health or the environment. Federal and state inspectors conduct both routine facility inspections and "for cause" use investigations. Evidence of misuse (e.g., samples, photos, statements and records) may be used to prosecute violators in federal or state jurisdictions (or in both) depending on the circumstances of the case. Penalties can be substantial. For example, FIFRA provides for a $6500 civil/administrative fine for each violation or count. In addition, criminal prosecutions are not unusual. While classified as misdemeanors, criminal offenses under FIFRA are considered serious environmental crimes and carry a maximum penalty of one year in jail per count. Two unlicensed pest control operators in Mississippi were sentenced to 5.5 and 6.5 years in a federal penitentiary. Sentences of 2 to 3 years for misuse of pesticides are commonplace, along with substantial fines. However, pesticide violations have decreased over the last two decades as education and knowledge of pesticide laws and regulations have become better known.

Good laboratory practices (GLP)

Working closely with the Office of Pesticides Programs, teams of investigators and scientists regularly conduct Good Laboratory Practices inspections at facilities that generate the scientific studies used in support of pesticide registrations. In addition, specific studies are randomly audited to verify adherence to identified protocols and procedures. Everything from the credentials of the researchers to the calibration of the equipment is thoroughly examined. The raw data are compared to the reported results to ensure accurate reporting. "For cause" audits of data are conducted when EPA scientists observe inconsistencies or irregularities in the studies submitted by the registrants.

A fair and vigorous enforcement program levels the playing field for the regulated community, removes any economic advantage of noncompliance (such as when using an unregistered pesticide on a site or crop not listed on the label) and exacts retribution as appropriate. As a result, enforcement is the exclamation point of the process that began with the registration of pesticides and the development of the labels and completes the mission of the EPA to provide a measure of consumer protection and to protect human health and the environment.

Summary

The US Environmental Protection Agency was formed in 1970 and became responsible for regulating the rapidly expanding development and use of pesticides. During the course of the next 20 years, the use of some pesticides was cancelled and testing requirements were developed to study the effects of pesticides on human health and the environment. These requirements are regularly revised to include the most recent developments in science. EPA toxicologists, chemists and biologists review proposed pesticide labels and revise label instructions as needed to ensure that human health and environmental safety will not be compromised. States may also register or approve pesticide labels for use in their jurisdictions and are allowed to add additional restrictions or requirements to the pesticide label. However, state guidelines cannot be less restrictive than those outlined on the federally approved label. The EPA and state regulatory agencies enforce pesticide laws regarding the purchase, use and disposal of pesticides. Pesticide labels are developed after years of research and include specific information about the pesticide and its use. The label is a legal document and all directions must be followed by those who use the product.

Photo and illustration credits

Page 145: Herbicide testing; William Haller, University of Florida Center for Aquatic and Invasive Plants

William T. Haller
University of Florida, Gainesville FL
whaller@ufl.edu

Appendix B

Aquatic Herbicide Application Methods

Introduction

All pesticide labels contain very specific information regarding how they are to be stored, handled and used. It is illegal to use any herbicide in, on or over water unless it is registered for that purpose and has aquatic use directions on the label. States may have pesticide use regulations that are more strict than federal regulations; thus, several states require that aquatic pesticide applicators be certified and licensed before they may purchase, handle and apply pesticides and that permits are obtained before pesticides are applied. Potential users of pesticides should contact state agencies such as county cooperative extension offices, state game and fish agencies or state environmental authorities to ensure compliance with any additional state-specific use restrictions.

A few herbicides may be applied directly from the container; for example, the labels of some copper sulfate herbicides suggest placing the dry granules in a cloth bag and towing the filled bag behind a boat to ensure uniform application throughout the water column. However, the majority of aquatic herbicides must be diluted or mixed with water before application. The purpose of the diluent (water) is to ensure consistent coverage of the target weeds so the herbicide can be absorbed into the plants. Most herbicide labels state that applicators should "use sufficient diluent to obtain uniform coverage of the target weed." Some labels are more restrictive and specify the amount of diluent to be used during application of the herbicide. For example, a label may specify "apply in 50 to 150 gallons of water per acre for adequate coverage." The public often believes that the mixture being applied to weeds is concentrated herbicide, but this is rarely—if ever—the case because herbicides are mixed with large volumes of water. Applicators are required by law to have the label at the application site and it is critical that they read the label carefully before aquatic herbicides are diluted, mixed and applied to ensure that the herbicide is applied in a legal, appropriate and effective manner.

Foliar applications

Foliar herbicides are mixed with water and sprayed on the foliage of floating or emergent plants in a given area. The goal during foliar application of an aquatic herbicide is to obtain good coverage and ensure that the maximum amount of herbicide is taken up by the target weed. Most floating and emergent plants have a waxy layer (cuticle) on their leaves and stems that must be penetrated in order for the herbicide to be taken up by the plant. The labels of some aquatic herbicides suggest or require the addition of surfactants that dissolve the cuticle and facilitate uptake of the herbicide by the plant. For example, a label may state that "a surfactant may be applied at a rate of 0.25 to 0.5% (1 to 2 quarts per 100 gallons) with the tank mix to get best results." In this example, the addition of a surfactant is not required by the label so its use is optional; other labels require the use of surfactants.

Just as carpenters and electricians have specialized equipment for their work, aquatic applicators often have tank- and pump-equipped boats and trucks for the application of herbicide treatments. A typical boat may hold a pump (calibrated to apply from 4 to 10 gallons per minute of a herbicide mix) and a 50- to 100-gallon mix tank. This equipment is calibrated to apply the correct amount of herbicide over the area to be treated. Selectivity, or the ability to control weeds growing among native plants, is usually accomplished by choosing the appropriate herbicide or by using a handgun to apply the herbicide mix only to the weeds and not to the desired native species. This is not always possible but is practiced as much as equipment and herbicide selection allow.

Most homeowners have small "pump-up" garden sprayers or backpack sprayers for lawn and garden use. Herbicide labels may include use directions for mixing the herbicide for small or localized spot treatments using small equipment. For example, if control of clumps of purple loosestrife along a shoreline is desired, the herbicide label may state "mix a 1 to 2% solution of herbicide in a backpack sprayer and spray weeds to wet." A gallon of water contains 128 fluid ounces, so the applicator would add 1.28 fluid ounces of herbicide to 127 fluid ounces of water to get a 1% solution. A 2% herbicide solution would be 2 x 1.28 fluid ounces, or 2.5 fluid ounces of herbicide per gallon of total tank mix. Be careful; some herbicides cannot be used in sprayers that will also be used for garden or ornamental plants, as some leftover herbicides can be quite toxic to other plants. Where is this information? On the label that is attached to every herbicide container!

The foliar application of herbicides to emergent and floating-leaved plants is generally well understood by homeowners because this is common practice on ornamental lawn and garden plants. The application of herbicides for submersed weed control, however, is often more complicated and thus more difficult to understand.

<u>Submersed aquatic applications</u>
The control of submersed aquatic weeds is much more difficult than control of emergent aquatic plants for the following reasons:

- Fewer herbicides are registered for submersed treatments
- The dilution effect of water depends on the depth of the water
- Wind, waves and currents dilute herbicides
- It takes more time to treat and cover submersed plants
- Submersed weeds are generally much more expensive to treat
- The growth stage and area covered by the plants are important
- Use of treated water for irrigation and drinking may be restricted

These general factors – and additional site-specific ones – determine which herbicides should be used to control submersed aquatic weeds. Water flow, dilution and water use are often the critical factors to consider when choosing a herbicide. Water flow and dilution may result in herbicide concentration/exposure times (CET) that are insufficient for herbicides to be effective (Chapter 11). There are also water restrictions on many herbicides for use in and adjacent to potable water intakes and for water used for irrigation. There are two general types of submersed aquatic weed applications, depending upon the CET requirements for the herbicides.

Contact herbicides

Contact herbicides are applied at relatively high concentrations, have very short half-lives in water and require a contact time of hours to a few days to kill plants. They include copper products, diquat, endothall and carfentrazone which may be applied along strips of shoreline and in relatively small areas where dilution is high, provided contact of the herbicide with the target weed is maintained for an amount of time sufficient to achieve control. The decision to use a contact herbicide is site-specific and the greatest chance of success occurs when herbicide applications are done on calm days to optimize contact times. Contact herbicides in general provide 3 to 6 months of weed control, depending upon the weed, geographical area of application (northern US vs. southern US) and length of growing season (Chapter 11).

Systemic or enzyme-inhibiting herbicides

Systemic enzyme-inhibiting herbicides are generally applied at concentrations lower than contact herbicides, must remain in contact with target weeds for relatively long times (up to 45 days or more) and are very slow to control submersed aquatic weeds. These herbicides are often applied as low-dose whole-lake treatments to control weeds throughout the lake. Systemic enzyme-inhibiting herbicides include fluridone, penoxsulam and imazamox. The former two herbicides are applied at rates of 5 to 20 ppb (parts per billion); concentrations can be maintained with additional treatments over several weeks to control hydrilla (Chapter 13.1), Eurasian watermilfoil (Chapter 13.2) and other submersed species. Imazamox is applied at 50 to 75 ppb and requires a contact time of several days. Penoxsulam and imazamox were registered in 2007 and 2008, respectively, and use patterns are still being developed (Chapter 11).

Systemic herbicides with short contact times

There are always exceptions to the rule, and 2,4–D and triclopyr are the exceptions in this case. Both are systemic herbicides but are absorbed in lethal doses by the target weeds in a relatively short time (1 to 4 days), depending upon the concentration applied. These two herbicides are effective for selective control of Eurasian watermilfoil and other dicot (non-grass) weeds. Concentrations of these herbicides for submersed weed control generally range from 1 to 2 ppm (parts per million). 2,4–D and triclopyr are applied at the highest labeled dose in areas where dilution is most likely to occur (such as small treatment areas and in strip treatments along shorelines) and on dense mature plants. Lower doses may be used in large treatment areas and in protected coves and bays with little water exchange.

Application of formulations

Herbicide formulation refers to how a herbicide is sold (as a liquid, granular or other form) and this determines the type of equipment needed for application of the herbicide. Many aquatic herbicides are sold as both liquid and granular formulations because many are used for both foliar and submersed aquatic weeds. For example, you would not apply 2,4–D as a granular formulation for foliar applications to purple loosestrife (Chapter 13.6); you would use a liquid formulation. The formulations of common aquatic herbicides are listed in Chapter 11.

Liquid formulations can be applied to submersed aquatic weeds in several ways, with the type of application determined by the specific location, size and depth of the treatment area. Surface applications are typically done along shorelines and under or around boathouses and docks where water depths average 3 to 6 feet deep. Granular and deep-hose applications are often used in deeper water, particularly in water where submersed weeds are growing in water from 6 to 20 feet deep. The objective of these deep-water treatments is to ensure that the herbicide mixes in the water column and reaches the plant beds where they can be taken up by the target weeds.

Effect of thermoclines

Temperature-dependent thermoclines often develop in lakes and other non-flowing waters during summer, particularly in northern regions. A thermocline occurs when the upper and lower portions of the water separate into warm and cool layers. Swimmers are often familiar with this phenomenon; for example, water in the upper layer of a lake feels warm, but diving down to depths of 6, 8 or 12 feet can be shockingly cold. This thermal stratification is well-known to applicators of aquatic herbicides as well and can reduce the effectiveness of herbicide treatments because the warm upper and cool lower layers of the water do not mix. Herbicides applied to the surface of the water may control upper portions of weeds, but treatments do not penetrate into the deeper cool layers. As a result, root crowns, rhizomes and low-growing plants below the thermocline are not controlled by the herbicide. The depth of the thermocline is influenced by water clarity and varies among lakes, but water temperature typically drops 2 °F for each 3' change in depth. If aquatic weeds are growing above and below the thermocline, deep-water injection of liquid herbicides or application of granular herbicides may be used to control weeds in both thermal zones.

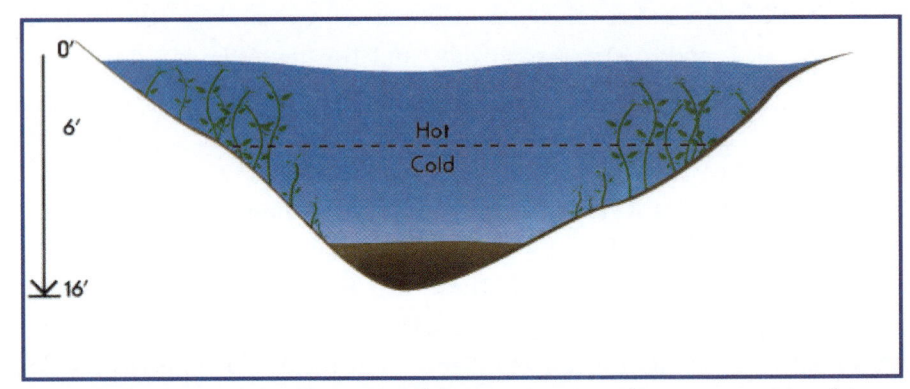

Foliar and submersed concentrations

The labels of glyphosate, 2,4–D, carfentrazone, triclopyr, diquat, endothall, copper, imazamox, imazapyr and penoxsulam products allow foliar applications for specific weed problems. Foliar-applied herbicides are usually mixed with 50 to 200 gallons of water per acre treated according to label directions and a surfactant is usually added to the tank mix to facilitate herbicide absorption or to ensure even coverage of the target plants. These herbicides are typically applied in "pounds per acre" with one pound of the herbicide's active ingredient in 100 gallons of water, resulting in a 0.1% concentration (1000 ppm). This relatively high concentration is needed to ensure that the plant absorbs enough herbicide to kill the weed on contact or through translocation to the site where the herbicide kills the plant.

Fortunately, application of herbicides for control of submersed aquatic weeds requires much lower concentrations of herbicides. This is because most submersed plants lack the waxy cuticles that slow herbicide uptake in many emergent plants and the leaves of many submersed plants are only a few cells thick. Tank mixes may still call for one pound of herbicide in 100 gallons of water, but in one acre-foot of water, the concentration of herbicide that contacts submersed plants is only 1/2.7 or 0.370 ppm (370 ppb) due to the dilution effect of the water being treated. Eurasian watermilfoil can be controlled with as little as 10 ppb of fluridone, but control of this weed with triclopyr or 2,4-D may require up to 2 ppm (2000 ppb). The ability of herbicides to control submersed weeds at such low concentrations contrasts sharply with the concentrations required to control larger, more tolerant floating and emergent weeds.

Although less herbicide is used per acre-foot of water for submersed weed control, submersed weeds often grow in water that is 8, 12 or 16 feet deep. Thus, submersed weed control often requires more herbicide per acre than foliar treatments due to increased water depth.

Selectivity

Weed control in an aquatic ecosystem is very different from weed control in an agricultural setting. For example, farmers want to control all the weeds in a cornfield without affecting the corn, whereas managers of natural and aquatic areas often wish to control a single weed species growing among 50 to 100 desirable native species. Research regarding selectivity of aquatic herbicides is ongoing and depends upon the following factors:

- Choice of herbicide: some herbicides control submersed weeds without affecting a number of other desirable nontarget plants, but the choice of herbicides that work in this manner is limited and complete selectivity is not always possible. As a result, herbicide selection is often dictated by the types of native species present in the proposed treatment area. In general, herbicides applied for submersed weed control have little effect on rooted emergent species due to the relatively low concentrations of herbicides used to control submersed weeds.

- Dose or amount of herbicide: not all plants are equally susceptible to herbicides. Application rates needed to control different weeds are usually listed on the herbicide label.

- Stage of plant growth: some herbicides used for submersed weed control can be applied in very early spring when weeds are actively growing and native plants are still dormant.

- Selective foliar application: handguns can be used to target and apply herbicides only to the weeds and minimize damage to nontarget species. However, this method is not feasible in most submersed treatments.

Although selective treatment of submersed weeds is more difficult than treatment of floating and emergent weeds, the reduction in growth and coverage of submersed weeds generally results in less weed competition and quick recovery of native species in the treated area. This occurs because most submersed weeds reproduce using vegetative means and many nontarget native plants reproduce by seeds. Elimination of dense weed canopies and the reduction of competition from invasive weeds often results in germination and growth of desirable species during the season of the herbicide treatment or soon thereafter.

Summary

Small-scale foliar application of herbicides to emergent and floating weeds is easily within the capabilities of most riparian homeowners, provided the correct herbicide is chosen and label directions are followed. The application of herbicides to aquatic weeds in large areas or for submersed weed control is more expensive, complicated and often requires specialized equipment to obtain the most cost-effective control. Selectivity results from a combination of factors, including herbicide choice, time of year and nontarget desirable species in the proposed treatment area. The size or area of the treatment site also affects the concentration-exposure time requirements for herbicides. In addition to label requirements, all these factors that affect submersed weed control clearly indicate that experienced state agencies responsible for permitting and managing aquatic resources be contacted prior to undertaking weed control projects.

For more information:
- How to build weighted trailing hoses. http://plants.ifas.ufl.edu/guide/building_weighted_trailing_hoses.html
- http://www.ecy.wa.gov/programs/wq/plants/management/aqua028.html
- http://aquat1.ifas.ufl.edu/guide/herbcons.html
- http://ohioline.osu.edu/a-fact/0015.html
- http://aquatplant.tamu.edu/index.htm
- University of Florida Center for Aquatic and Invasive Plants. http://plants.ifas.ufl.edu

Photo and illustration credits
Page 152: Herbicide application; William Haller, University of Florida Center for Aquatic and Invasive Plants
Page 154 upper: Herbicide application; William Haller, University of Florida Center for Aquatic and Invasive Plants
Page 154 lower: Thermocline; Joshua Huey, University of Florida Center for Aquatic and Invasive Plants
Page 155: Submersed application; Lyn Gettys and Joshua Huey, University of Florida Center for Aquatic and Invasive Plants

Appendix C

Bernalyn D. McGaughey
Compliance Services International
bmcgaughey@complianceservices.com

A Discussion to Address Your Concerns: Will Herbicides Hurt Me or My Lake?

1. Our lake is pristine and we don't want to put dangerous chemicals in it. Why should we use herbicides now?

A pristine lake is balanced, stable… and very rare, especially when lakes are surrounded by homes or used for recreation. The lakes we live near and play in are often inundated by excess nutrients and foreign and invasive species. Most water bodies that require herbicide treatment have experienced explosive growth of invasive aquatic plants. While your lake may seem natural and pristine, there are sufficient nutrients in the water to allow exotic weeds – which don't belong in the lake – to dominate the system. Control of these weeds will enhance plant diversity and water quality (both of which are degraded by dense weed growth) and will help restore the overall health of the lake.

Your lake association or responsible public agency has evaluated all the options for aquatic plant management and has decided that the most effective means of controlling weeds at this point is to use herbicides. The herbicides that will be used are biodegradable and will not affect the pristine nature of the lake in the long term. When used by professionals according to label directions, herbicides are not "dangerous chemicals" but instead are curative products that have been extensively tested and can effectively control nuisance and invasive aquatic weeds.

2. How dangerous are these chemicals? How do we know they're safe?

Interestingly, aquatic herbicides are one of the smallest niches of specialty weed control products (Chapter 11), yet they are also among the most extensively tested. Because these products are added directly to water, the EPA requires extensive data to assess the safety of a herbicide before it can be registered for use in aquatic systems (Appendix A). Many years of testing and use have shown that registered aquatic weed control products can be used safely in all areas of the US. In addition, many years of safety and monitoring tests in the laboratory and in the field have been conducted to determine exactly how a given product should be used in a particular situation. It is also important to remember that the treatment level (or concentration in water) of a herbicide is typically much lower (100- to 1000-fold more dilute) than the concentration that might be harmful to you, your pets or nontarget organisms that live in the lake.

The data required by the EPA for registration of an aquatic herbicide are generated in studies that are conducted according to stringent protocols of conduct, design and evaluation. For example, a single study is conducted using a testing guideline that describes the number of organisms that must be tested, how they are housed and even the temperature and daylength under which the organisms must be maintained. The test is also governed by a series of "Standard Operating Procedures" that have additional parameters for testing and documentation. The guidelines for the test are further supported by a "Standard Evaluation Procedure", which outlines the criteria that must be met in order for the study to be defined as "acceptable." The EPA toxicologist produces a "data evaluation record" for the study and ultimately classifies the study as acceptable or unacceptable for incorporation into the risk assessment process. In a parallel requirement, the Standard Operating Procedures mentioned above must be conducted following formal "Good Laboratory Practice" requirements as outlined by the EPA. Good Laboratory Practice Standards are validated through both internal and external audits. Once a study has conformed to all of the requirements for study acceptance, data generated by the study are combined with data from all other acceptable studies of the herbicide and a risk assessment profile is developed.

The risk assessment process is complex and requires identifying which studies should be integrated into the hazard and exposure evaluation process. The 84 to 124 different studies required for registration of an aquatic herbicide take from 6 to 10 years to complete and are integrated in a robust scientific assessment that is evaluated by the EPA in a process that can take an additional one to three years before labels are approved.

3. Do these herbicides break down in the environment? I realize the herbicides themselves have been evaluated by regulatory agencies but what about their breakdown products?

Identification and evaluation of the components into which an herbicide breaks down is a critical and required part of the data that must be submitted as part of a product's registration process. Degradation and metabolism pathways must be studied and the molecules that are produced along those pathways must be identified. If any molecules are believed to be "of toxicological concern" (and there is a definition for that), then those molecules must be tested as well, both alone and in combination with the original or "parent" molecule.

Testing of breakdown products is not limited simply to toxicity; breakdown products must also be evaluated for their persistence in the environment. In addition, the mechanism (light, heat, microbial action) that produces them and acts to further break them down must also be understood. The final fate of the parent and breakdown products must be completely identified, reported and understood by chemists and toxicologists. Additionally, there are flagging criteria that are used to put "stop lights" on certain uses or environmental introductions of herbicides. These "stop lights" can be associated with direct toxicity, persistence, bioaccumulation or other important environmental and toxicological properties of the pesticide. If a product is flagged by one of these "stop lights" during testing, the company developing the product (especially one that will be used in water) may reconsider whether to proceed with the high cost of registration if there is a good chance the product will not successfully make it through the registration process.

4. If the chemical companies do the research and submit their data to the EPA, isn't this like the fox guarding the henhouse? Their data may be falsified!

With the current regulatory standards and rigor of EPA review, it is virtually impossible to falsify the data supporting a product. Companies submitting studies must certify that they are conducted in accordance with EPA regulations for good laboratory practices and usually hire independent quality assurance scientists to conduct audits as the studies are performed. In addition, the EPA has established a random laboratory and study audit program. This program has the authority to audit laboratories that conduct studies in support of pesticide registration and companies that sponsor them and can randomly select submitted studies for auditing. It must be possible during this audit process to confidently recreate the entire study from the "raw data" (laboratories are legally required to maintain all data for any submitted study on which registration relies). If a problem is found or the results cannot be reconstructed, not only is the study rejected for regulatory use, but the facility conducting it or the company sponsoring it is likely to undergo a more complete audit of all studies conducted during the same period, at the same facility or on the same product. Penalties for falsifying studies can be severe and include fines and/or imprisonment.

5. If herbicides make up only part of the chemicals that are applied, how do we know whether any other part of the product or its inert ingredients are dangerous?

First of all, let's understand a little bit about herbicide formulations. The chemical that controls the weed, in its pure form, is called the "active ingredient." The *technical grade* of the active ingredient is used in testing, and that technical grade must contain all those components that are found in the typical manufactured product that makes up the active ingredient. Technical grade chemicals are usually very pure (98%+), but may include additional compounds that are formed as the active ingredient is made. Components in the technical grade product, other than the pure active ingredient, are usually remnants of the manufacturing process, molecules that are impossible to separate from the parent compound, or other unintentionally added ingredients. All such impurities must be identified even if they are present in extremely low quantities. If any are of toxicological concern, they must be removed from the technical product or reduced to levels considered acceptable by the EPA.

Testing with the technical grade of the herbicide will identify toxic and environmental effects that might be caused by the active ingredient itself or any chemical components formed by the active ingredient. The technical grade form of herbicides are too concentrated and are rarely useable as herbicides without some modification to allow proper measurement (dilution by water, clay granules or other solvents or carriers), tank mixing (conditioners, such as emulsifiers, anti-foaming agents or wetting agents), and stability and distribution to the target site (by use of surfactants, drift control agents, dyes or other similar agents). The proper addition of these materials to the technical grade product produces an *end use formulation*, which is what is then purchased and used in weed control. This end use formulation must also be tested, but in a limited way <u>unless</u> the initial tests show that there is a measurable difference in toxicity between the *technical product* and its *end use formulation*. If there is a difference, the typical remedy is to change the components of the formulation so that they do <u>not</u> affect the toxicity or environmental characteristics of the end-use formulation.

Collectively, the formulation products discussed above are often referred to as "inert ingredients" because they do not contribute to the activity of the active ingredient. Formulations are considered trade secrets because their components may provide a competitive advantage and will be associated with a brand trademark. As such, the "secrecy" surrounding inert ingredients is one of competition, not toxicological properties. Additionally, not just any compound can be used in a formulation. The EPA requires that all inert ingredients in pesticide products be cleared prior to use and maintains a list of products from which the formulation chemist can choose. If the formulation chemist chooses a product that is not on the cleared list of inert ingredients, then supporting data must also be submitted for that "inert" ingredient. A separate and thorough review process will determine whether the inert ingredient can be added to the EPA's cleared list and safely used in the subject formulation. Incidentally, these inert ingredients are not "secret" from the EPA. Each technical and end use product must be supported by a complete "confidential statement of formula" so that the EPA can evaluate the acceptability of the full product and its additives. The confidential statement of formula is also used by the EPA when random or purposeful samples of the product are pulled from chemical plant distributors or applicators and analyzed for their compliance to the stated formula.

Inert ingredients in products to be used on food (and most aquatic uses are considered food uses due to the subsequent exposure to fish and shellfish, which in turn could be food items for people) or potable water must also have tolerances (allowable dietary levels of the product and any breakdown products of concern) set under the Federal Food, Drug and Cosmetic Act, which is administered by the Food and Drug Administration. Scrutiny of products that are used in, or may reach drinking water sources, is especially intense because the underlying assumption is that exposure could occur over a lifetime, from any and every drinking water source. In the case of aquatic herbicides, this assessment process greatly overstates exposure and thus results in a very conservative risk assessment.

6. When will it be safe for my kids to swim in the water again?

Each herbicide has a specific label statement regarding water use and swimming after weed treatment. Label statements are based on the results of various studies and the risk assessment process described above. Swimming restrictions listed on the label are most often related to the dissipation of the herbicide in water and added "safety factors" that build in at least a 100- to 1000-fold margin between what is observed in studies as a "no effect level" and the potential exposure level when a lake is treated. Therefore, the restriction interval (if any) is related to all studies conducted on the degradation and dissipation of the product and its dermal, oral and dietary toxicity, as well as any potential to irritate the skin or eyes or penetrate the skin. Herbicides that lack swimming restrictions may dissipate very quickly and/or the toxicity of the product at treatment levels is far below the "no effect level" in studies supporting product registration.

7. Will herbicide treatments kill the fish in our lake?

Aquatic herbicides are extensively tested for their effects on fish and other nontarget aquatic organisms. For the most part, these products are relatively non-toxic to fish because their mode of action (the way they affect the target weed) is based on photosynthesis or other plant processes that differ from animal biochemistry. A few types of aquatic herbicides (usually algicides) are toxic to fish at or near treatment levels, but application techniques that provide fish with the opportunity to escape from treated waters can reduce or prevent the loss of fish populations. This information is on the herbicide label; applicators are required to read and follow all label directions and precautions.

The applicator must consider the amount of plant cover and the manner in which it will be treated in his professional assessment of the needs of the lake. Decomposing vegetation can deplete oxygen levels in water, which can cause fish mortality if application precautions are not taken. Extreme infestations of weeds may require treatment of the lake in stages instead of using a single whole-lake treatment. Partial treatment will allow fish to escape to untreated, oxygenated waters as target plants in the treated area decompose.

8. The herbicide label says that the product is "toxic to fish and wildlife." Does this mean the herbicide treatment will kill our fish? If not, why do these chemicals kill plants without harming people or fish?

The statement referenced here historically has been required on a label when a pesticide intended for outdoor use contains an active ingredient with a fish LC$_{50}$ (acute toxicity level) of less than 1 ppm [equal to one part (or molecule) herbicide per one million parts (or molecules) of water]. "LC$_{50}$" is an abbreviation for "lethal concentration 50%" and represents the calculated concentration of the substance that is expected to kill 50% of the organisms studied. The standard label statement required in this case is, "This pesticide is toxic to [fish] [fish and aquatic invertebrates] [oysters/shrimp] or [fish, aquatic invertebrates, oysters and shrimp]." Likewise, if the product "triggers" a toxicity level preset for birds or mammals, a similar statement is required. When a pesticide intended for outdoor use contains an active ingredient which has a mammalian acute oral toxicity of less than 100 mg material/kg bodyweight, an avian acute oral toxicity of less than 100 mg/kg, or a subacute dietary toxicity of less than 500 ppm (500 parts of material per 1,000,000 parts diet, by weight), the label must state "This pesticide is toxic to [birds] [mammals] or [birds and mammals]." It is important to note that pesticides with lower LC$_{50}$ values are more toxic than those with higher values. For example, a product with a toxicity of 100 mg/kg is more toxic than one with a toxicity of 250 mg/kg.

There are several circumstances that can make toxicity to organisms in the field less severe than suggested by the label statement when herbicides are used for weed treatment. Some of these are:

Effective control levels: most aquatic herbicides are applied at rates well below those that would cause fish or wildlife toxicity. This is either because the target weed is particularly sensitive to the herbicide or because the herbicide interrupts a biochemical pathway that animals do not possess.

Application techniques: your professional applicator or supervising state agency knows what precautions to take for products that have a treatment rate close to a wildlife effect level. These precautions can include partial lake treatments; optimal treatment timing at the lowest rate possible; the use of drift control agents; and other informed choices made by the professional applicator.

Dissipation rate: Some aquatic herbicides essentially break down immediately or are rapidly absorbed by plants and vegetative matter. Studies to determine fish toxicity are conducted in pure-water systems (without plants) over a period of several days. Such studies provide comparable standards for judging toxicity and regulating products, but they are not necessarily equal to fish exposure and product toxicity in a natural, living system when an herbicide is used according to label directions.

Sediment binding: Some aquatic herbicides ultimately bind to organic matter, algae and soil particles and partially end up in lake sediments, where they may be metabolized by microbes or made unavailable through the physical process of mineralization. A product that is bound in the soil this way rarely presents a toxicity concern.

9. Is it safe to eat fish from the lake after herbicides have been applied?

No aquatic herbicides currently registered by the EPA have fish consumption restrictions. There are no restrictions because herbicides have established "tolerances" that are set by the EPA and the FDA. Tolerances are boundaries for acceptable levels of pesticide residues in food and are established after review of submitted data and in accordance with the Federal Food, Drug and Cosmetic Act. If an aquatic herbicide has tolerances set for fish, then the label will instruct whether the fish can be consumed immediately after treatment or if there is a waiting period. Where there is no established tolerance (either because the registrant has not sought it or due to the properties of the product), the label will prohibit the consumption of fish from a treated lake until enough time has passed for no residues of the product to be found in fish tissues. Professional applicators are well aware of the restrictions necessary for fishing and fish consumption, as these restrictions are clearly specified on the herbicide label. Applicators are required to post signs or otherwise clearly inform lake users of any water use restrictions.

10. How long does it take for herbicides to break down? Do the chemicals become concentrated in the fish or the sediment of the lake?

There are some specialized terms that will help you understand the metabolic processes that are at the root of this question. They are *adsorption, depuration, bioaccumulation* and *bioconcentration*. *Adsorption* is the manner and rate at which an organism assimilates a chemical into its system, whereas *depuration* is the manner and rate at which the organism rids itself of a chemical. *Bioaccumulation* occurs when the rate of adsorption (taking up the chemical) exceeds the rate of depuration (ridding of the chemical) during the period of exposure. When exposure is stopped, depuration continues and the organism will gradually clear itself of the chemical. Some scientists debate whether there is a difference between bioaccumulation and bioconcentration. However, *bioconcentration* is slightly different than bioaccumulation because the levels of a chemical that bioconcentrates build up and become more concentrated over time. This occurs because depuration is non-existent or very slow, so the organism never clears the chemical from its system and may build up higher and higher concentrations upon every exposure to that chemical. Bioconcentration does not occur in any currently registered aquatic herbicide. A herbicide may have a short bioaccumulation period in edible organisms like fish and in such a circumstance would be labeled with restrictions to prevent consumption until the depuration process has cleared the chemical from the organism's system.

Some aquatic herbicides may accumulate in sediments, but as discussed above, this is typically also associated with sediment binding that limits the biological availability of the product. The EPA takes into account potential accumulation of pesticides in fish and sediment prior to registering any product for use in water. In fact, pesticide accumulation in living systems or the environment is one of the "stop lights" discussed in Question 3 above. It is unlikely that any chemical that bioconcentrates would be registered for outdoor use in today's regulatory environment. It is possible that a product that bioaccumulates might be registered, because in most instances this property can be managed by restricting application rates, treatment intervals and consumption of treated organisms. If risks to man or the environment are unacceptable or unmanageable, then the product simply will not be registered.

11. Are aquatic herbicides carcinogens? Will they give me cancer?

There are currently no registered aquatic chemicals that are classified as carcinogens. The treatment of water systems with herbicides is considered a widespread use with high potential for human and nontarget organism exposure. Consequently, products registered for use in water must present a very low risk profile, even when – in the case of aquatic herbicides – potential exposure to humans is neither pervasive nor long term. Any legitimate evidence of carcinogenicity would immediately put the registration and use of an aquatic herbicide in jeopardy.

This brings up an area that confuses many people – how to interpret different kinds of studies with respect to their validity for use in the "risk equation." A number of factors contribute to the validity of a study, such as the purity and reliability of the test system (contaminants not found in the product or nature, or the use of unusual species or strains of test animals that could create false results), the statistical power of the experiment itself (inadequate numbers of test organisms or improper statistical analysis of results could yield false conclusions), or the route of exposure (an exposure route impossible in nature, such as intravenous injection of high concentrations of chemical). For these reasons, some studies are not used in the risk assessment process, provided there is a body of reliable information that contradicts their findings. In the event a new finding is of concern, the EPA has the means to restrict use, cancel use or put other protective measures in place until additional data are generated or assessed.

12. Plants that have been treated with herbicides rot and sink to the bottom of the lake and cause a buildup of muck. We don't want muck buildup so we shouldn't use herbicides, right?

The best time to treat with herbicides is usually in the spring when plants are very actively growing but still small. This practice results in very insignificant organic matter additions to the lake. Furthermore, research has shown that when the growth of plants is restricted or controlled with herbicides or other means, much less organic matter is produced than if plants are left untreated. Plants that are not managed in some way grow until they reach their full annual biomass and then naturally die back each winter; as a result, all the material produced by a plant over the course of the year is added to the lake annually. By reducing plant growth, herbicide use can actually reduce organic matter production and accumulation. Another factor contributing to "muck" is sedimentation. Dense stands of weeds tend to trap particles suspended in the water column and increase sedimentation or "muck" buildup.

13. I've watched herbicide applications in other lakes and the applicators always wear "moon suits" and all sorts of protective gear even though the label says we can swim and fish immediately after application of the herbicide. This makes no sense – what gives?

Pesticide labels are developed to take into consideration both the exposure to workers (handlers and applicators) and the exposure to the environment. Workers repeatedly handle concentrated herbicides before they are diluted for application. Therefore, applicators are required to wear personal protective equipment to minimize their exposure to high doses of chemical if the chemical properties of the concentrated herbicide pose a risk to them. Herbicides are diluted literally millions of times when they are applied to water and they are usually applied once per season. As a result, the same precautions are simply not necessary for any lake water users who are not repeatedly exposed to high concentrations of herbicides. For comparison, a tablespoon of salt in a batch of yeast dough contributes to the flavor and perfection of the final loaves of bread – but a tablespoon of salt taken alone could be dangerous for you.

14. People used to say that DDT, chlordane and all those other pesticides were safe and now they're banned. Will this happen with more modern herbicides too?

DDT was first registered as a pesticide in the 1940s; chlordane was first registered in 1948. Both of these compounds were insecticides and are in no way related to any currently registered aquatic herbicides. There is absolutely no comparison to the testing standards and regulatory requirements in place today with the meager parameters that were in place in the first half of the last century. Needless to say, our understanding of science, toxicology and the environment has increased tremendously in the last 50 years.

The oldest registered aquatic herbicide appeared first in the late 1950s. Any products surviving since then have been subjected to additional reviews and many additional data requirements, culminating in updated and more rigorous risk assessments, including reregistration. It is a testimony to their safety that, as testing and registration requirements increase, older aquatic herbicides are still in use today. In fact, with the additional testing, many restrictions have actually been *removed* from older products. Products developed over the course of the last 30 years, during our cycle of increased understanding and advanced science, are designed to have a minimal impact on the environment and are simply not comparable to the "first generation" pesticides like DDT and chlordane. Today's products are developed with the knowledge of their toxicity and impact and would not be registered or commercially developed if they carried a high "risk burden."

15. I agree that we have to use herbicides to get our weed problem under control, but how can we as residents reduce the risks associated with the use of these chemicals?

First of all, by taking the time to read and understand this manual, you have already invested in reducing your own risks, because you now understand the importance of following label directions and the instructions provided to you by your professional applicator.

Second, plan carefully and completely for a herbicide application in the early stages of an aquatic weed infestation so that your lake can be treated at the optimum time of the year with the lowest effective treatment rates, which can reduce the need for multiple treatments. This action will likely provide more effective weed control, reduce costs and lower the total amount of chemical that may be required for adequate weed control.

Additionally, many states have regulatory agencies that conduct additional risk assessments to refine their understanding of product properties as specifically as possible for the conditions in their state. In some cases, specific permits or precautions are required on a treatment-by-treatment basis, thereby further ensuring that lake residents and users understand the restrictions, if any, on the use of the lake or its resources. For example, New York takes an additional precautionary step and adds another layer of protection by restricting swimming in any treated lake for 24 hours after any pesticide application to its waters – even though scientific data, the label and product properties do not call for this additional precaution.

The risk-reducing protections necessary for safe use of a registered product are already in place once the product is registered. All you have to do is follow the label, the instructions of the applicator and any additional local regulations.

16. What exactly is risk? I don't want any risk!

We cannot live in a risk-free environment. Living near a lake is in itself a "risk." Risk, as related to the science of risk assessment, is poorly understood by anyone other than risk-assessment scientists. Most people equate "risk" with "being exposed to a risk", but these are not the same thing. Risk assessors deal with the likelihood (or probability) of an event happening <u>at all</u>, while being <u>at risk</u> is the likelihood of being affected by an event that is known to happen. Thus, the risk assessor will come to a conclusion (for example) that a given dose of a chemical has a <u>one in a million</u> chance of **causing** cancer, while the statistician following causes of death will report that an individual has approximately a <u>one in four</u> chance of **dying** from cancer. Two very different endpoints.

When we put actual quantifiable risks in perspective, the risk of harm from an aquatic herbicide (or any pesticide, for that matter) is negligible. The National Safety Council (2005) reports the following:

- The leading causes of death in the US are heart disease, cancer, stroke, respiratory disease and unintentional accidents, in that order.
- Of unintentional accidents, the fourth ranked cause of death is drowning. The odds of drowning in natural water (as opposed to a swimming pool) are 1 in 2,378.

No risk estimate for the effects that <u>might</u> result from exposure to a pesticide even begins to approach this number.

In risk assessment, the end point sought is that the **probability** of a risk is so low that it is expected to not occur. In risk assessment, "risk" is defined as the relationship between hazard and the likelihood of exposure. When aquatic herbicides are used in a lake, most residents and lake users will have little or no exposure to the product used for weed treatment, based on the application methods, precautions taken and infrequency of treatment. Your risk of suffering from an event related to herbicide use and exposure is miniscule.

17. Does the EPA guarantee that these herbicides are safe?

The regulatory language of FIFRA (Appendix A) actually prohibits descriptive language that would imply any registered pesticide is "safe." In part, this is because "safe" is a relative term that could easily be misleading. No agent, natural or man-made, is completely "safe." Even water, which is essential for life, can be dangerous if too much is consumed because in excess it can disrupt the balance of electrolytes in a living system. Electrolyte imbalance can lead to shock and eventual death if not corrected.

As discussed above, EPA registration requirements and the risk assessment process supporting a pesticide registration are intense and thorough. The directions for use that are listed on the product label take into account risk management measures that are necessary to reduce the risk of exposure to the point where there is no reasonable expectation of environmental or human health effects. Furthermore, there is now a revolving and formal Registration Review process, assuring that new scientific procedures and risk assessment methods are applied through a revolving process to all EPA registered products over the life of their registration.

18. Who else studies these chemicals besides the EPA?

Chemical use and its effects on the environment are closely scrutinized by many groups, including independent university scientists, state regulatory agencies, environmental groups and even the chemical companies themselves. Additionally, as the world economic and regulatory systems become more global, there is a closer coordination between countries in their requirements for and review of data on chemicals.

There are also protections written into FIFRA with respect to the discovery of previously unobserved effects. If a legitimate finding is made known to the company holding the registration for the chemical, that company must, within 15 days, report that finding and its significance to the EPA. If the EPA deems that the event is critical, it can immediately stop the sale or otherwise limit the use of the product. If the significance of the event is not major, but requires further understanding, the EPA may issue additional data requirements so that the initial finding can be studied and causes for it can be determined. Failure to follow these reporting requirements carries heavy penalties.

19. Big corporations are only interested in making money – they don't care whether their product is safe!

The development, registration and marketing of a pesticide take place in a highly visible segment of business in which relatively few companies compete. Add to that the extra burden of registering products for use in water systems and the general business risk couldn't get much higher. This is a mature industry with extremely high standards, a heavy regulatory obligation and a tremendous amount of exposure. Corporations employ scientists to conduct the research required for pesticide regulation, and these scientists eat the same food and use the same resources that we all enjoy. No company in such an environment would survive negligence, data falsification or poor business ethics. The mistakes of the early years that occurred in an emerging regulatory system and a budding scientific understanding of the environment that surrounds us are simply not inherent to the business today. They are of the past. Today's aquatic herbicide registrants are heavily invested in the safe and beneficial use of their products, environmental stewardship and sustainable practices. They have to be, or they wouldn't be here tomorrow. And being here tomorrow is how they survive, not simply by making money with no future in sight.

Appendix D

John D. Madsen
Mississippi State University, Mississippi State MS
jmadsen@gri.msstate.edu

Developing a Lake Management Plan

Introduction

Invasive aquatic plants are a major problem for the management of water resources in the United States. Nonnative invasive species cause most of the nuisance problems in larger waterways and often produce widespread dense beds that obstruct navigation, recreation, fishing and swimming and interfere with hydropower generation. In addition, dense nuisance plants increase the likelihood of flooding and aid in the spread of insect-borne diseases. Invasive plants also reduce both water quality and property values for shoreline owners.

Invasive species have a negative impact on the ecological properties of the water resource. They may degrade water quality and reduce species diversity while suppressing the growth of desirable native plants. Invasive species may alter the predator/prey relationship between game fish and their forage base, which results in higher populations of small game fish. Invasive species may also change ecosystem services of water resources by altering nutrient cycling patterns and sedimentation rates and by increasing internal loading of nutrients.

The most troublesome invasive plants that cause problems in the United States are listed in the following table. These species and recommendations for managing them are discussed in Chapter 13 of this manual. These exotic weeds are most likely to cause the greatest concerns, but many other native and nonnative species can cause problems as well, particularly in small areas or in ponds.

SUBMERSED		
Common name	Scientific name	Described in:
Hydrilla	*Hydrilla verticillata*	Chapter 13.1
Eurasian watermilfoil	*Myriophyllum spicatum*	Chapter 13.2
Curlyleaf pondweed	*Potamogeton crispus*	Chapter 13.7
Egeria	*Egeria densa*	Chapter 13.8

EMERGENT		
Common name	Scientific name	Described in:
Waterchestnut	*Trapa natans*	Chapter 13.3
Purple loosestrife	*Lythrum salicaria*	Chapter 13.6
Phragmites	*Phragmites australis*	Chapter 13.9
Flowering rush	*Butomus umbellatus*	Chapter 13.10

FLOATING		
Common name	Scientific name	Described in:
Giant and common salvinia	*Salvinia molesta, S. minima*	Chapter 13.4
Waterhyacinth	*Eichhornia crassipes*	Chapter 13.5

Development of a management plan

Water resource managers need to have an aquatic plant management plan for long-tem management, even in bodies of water that have not yet been invaded by these exotic species. An effective aquatic plant management plan should establish protocols to prevent the introduction of nuisance plants, provide an early detection and rapid response program for the waterbody so new introductions can be managed quickly at minimal cost and aid in identifying problems at an early stage. The plan should also assist in identifying resources and stakeholders so that coalitions can be built to aid in the management of problem species. The planning process should include information that is already available and identify gaps in knowledge where more information is needed. An effective management plan will help water resource managers communicate the need for management of invasive species and provide a rationale or approach for management. A comprehensive aquatic plant management plan should have eight components: prevention, problem assessment, project management, monitoring, education, management goals, site-specific management and evaluation.

Prevention

The focus of a prevention program is education and quarantine combined with proactive management of new infestations (early detection and rapid response). Most invasive aquatic plants are introduced to a water body as a result of human activity and introductions most often occur when invasive plants are transported on boats, watercraft and boat trailers. Prevention activities can include signage at boat launches and marinas and other educational programs. Successful prevention programs utilize federal and state legislation, enforcement, educational programs in broadcast and print media and volunteer monitoring programs. An early detection and rapid response program should be employed in conjunction with prevention efforts to control new infestations at an early stage. Proactively controlling new infestations before they develop into large populations of exotic plants is both technically easier and less expensive, which results in major cost savings in the long run. The eradication of small populations is much more likely than eradication of large established populations. Early detection and rapid response is a critical component of an exotic species prevention program and is emphasized by federal agencies involved in invasive species management.

Problem assessment

Problem assessment should focus on identifying a problem in a given waterbody and collecting information about the problem. This information can then be used to formulate specific problem statements that define the cause of the problem. Problem assessment is the process of both acquiring objective information about the problem, such as maps and data on plant distribution, and identifying groups or stakeholders that should have input into formulating the problem statement. Problem assessments should also identify the causes of the problem and should increase the understanding of the water resource by reviewing information that is already available and highlighting areas where additional information is needed. A specific problem statement should be developed using the resources identified during problem assessment to aid in refining the concerns of users and the nature of the nuisance problem.

Project management

Project management is often a neglected aspect of managing invasive plants, particularly when volunteers manage the project. Successful projects are the result of good planning and management of assets, which include financial resources, partnerships, volunteers and other personnel. Detailed records of expenses must be maintained, particularly if the project is funded by government entities.

In addition, a thorough evaluation of success of the program should include expenditures of both time and labor.

Monitoring

A monitoring program should include not only an assessment of the distribution of the target plant species, but also a program to monitor other biological communities (including desirable native plant communities) in the water body. Water quality parameters should be recorded on a regular basis to determine whether long-term changes have taken place in the water body and to assess whether management activities have had a positive or negative effect on other aspects of the water resource. Monitoring should also include baseline data collection (as outlined in the problem assessment section above), compliance monitoring involving a permit and assessments of management impacts to the environment at large. Successful monitoring programs often include a "citizen" monitoring component. For instance, citizen monitors have assessed water quality in many water bodies for several decades using techniques as simple as measuring water clarity using a Secchi disk. The largest volunteer network in the US is The Secchi Dip-In (http://dipin.kent.edu/secchi.htm), though many states also have a statewide volunteer network (e.g., Florida, http://lakewatch.ifas.ufl.edu/; Maine, http://www.mainevolunteerlakemonitors.org/waterquality/; and others).

Education and outreach

Education and outreach should be initiated at the beginning of the program and should continue throughout the project. Education initially consists of familiarizing the project group with the problem and possible solutions, which helps to build a consensus regarding the solution. As the program progresses, education efforts should be extended to include the public (in addition to stakeholders in the lake association) and to inform them of the problem, possible solutions and what actions the program is taking to address the problem. It is important to provide as much information as possible to the public and to be forthright and open about management activities. A public web page devoted to the management program can be a very successful tool but the project group should utilize local media outlets, such as newspapers and radio, as well. Also, if your project is successful, share your success with others through homeowners associations or your local county cooperative extension service.

Plant information and methods

The development of a program to monitor invasive plants requires a list of invasive, nonnative, native, endangered and threatened plant species in the waterbody, maps marked with the locations of species of concern or species targeted for management, locations of nuisance growth and bathymetric maps. Quantitative plant data (sampling for plant distribution or abundance using a recognized sampling protocol) should be used for assessment, monitoring and evaluation as often as possible. Quantitative data is more desirable than qualitative data (subjective assessments such as "a big population" or "heavily infested") because:

- Quantitative data is objective and provides hard evidence regarding the distribution and abundance of plants, whereas subjective surveys are based on opinion rather than fact
- Quantitative data allows for rigorous statistical evaluation of plant trends in assessment, monitoring and evaluation
- Quantitative data and surveys may eliminate costly but ineffective techniques in a given management approach
- Quantitative data allows individuals other than the observer to evaluate the data and to develop their own conclusions based on assessment, monitoring and evaluation data

Plant quantification techniques vary in their purpose, scale and intensity (see table below). Cover techniques include both point and line intercept methods. These techniques yield the most information regarding species diversity and distribution and can reveal small changes in plant community composition. The best method for measuring plant abundance remains biomass measurement but this is time-intensive and usually reserved to evaluate the effectiveness of management activities. Hydroacoustic surveys measure submersed plant canopies while the plants are still underwater and are excellent for assessing the underwater distribution and abundance of submersed plants; however, this technique is unable to discriminate among species. Visual remote sensing techniques, whether from aircraft or satellite, have also been widely used to map topped-out submersed plants or floating and emergent plants.

Aquatic Plant Quantification Techniques	
Technique	Information produced
Cover techniques: point intercept	Species composition and distribution (whole-lake)
Cover techniques: line intercept	Species composition and distribution (study plot)
Abundance techniques: biomass	Species composition and abundance
Hydroacoustic techniques: SAVEWS	Distribution and abundance (no discrimination among species)
Remote sensing: satellite, aircraft	Distribution (plants near the surface only, no discrimination among species)

Management goals

Specific management goals that are reasonable and testable should be formulated as part of the management plan. This set of goals provides the milestones that can be used to determine whether the management program is successful. If specific management goals are not established, stakeholders may dispute whether management efforts have been successful since they may lack a clear understanding of the expectations of the management program. Goals should be as specific as possible, including indicating areas that have a higher management priority.

Providing stakeholders with a specific set of goals will allow them to evaluate quantitative data to determine whether management goals have been met. For instance, if vegetation obstructs recreational use of the waterbody, a goal of "unobstructed navigation" is vague and may result in unending management. If, however, the goal is to maintain navigation channels in navigable condition 90% of the time, then the success of the management program can be measured, tested and compared to the specific goal. Once plant management goals are developed, methods to achieve the goals should be implemented using techniques that are acceptable to stakeholders and regulatory agencies based on environmental, economic and efficiency standards. Management techniques will vary based on conditions within the water body and frequently change over time; this is referred to as site-specific management.

Site-specific management

Site-specific management utilizes management techniques that are selected based on their technical merits and are suited to the needs of a particular location at a particular point in time. Techniques should be selected based on the priority of the site, environmental and regulatory constraints of the site and the potential of the technique to control plants under the site's particular conditions.

Spatial selection criteria include the identity of the target weed species, the density of the weed, the size of the infested area, water flow characteristics, other uses of the area and potential conflicts between water use and restrictions associated with selected management techniques. For example, consider an area of nuisance growth that is close to a drinking

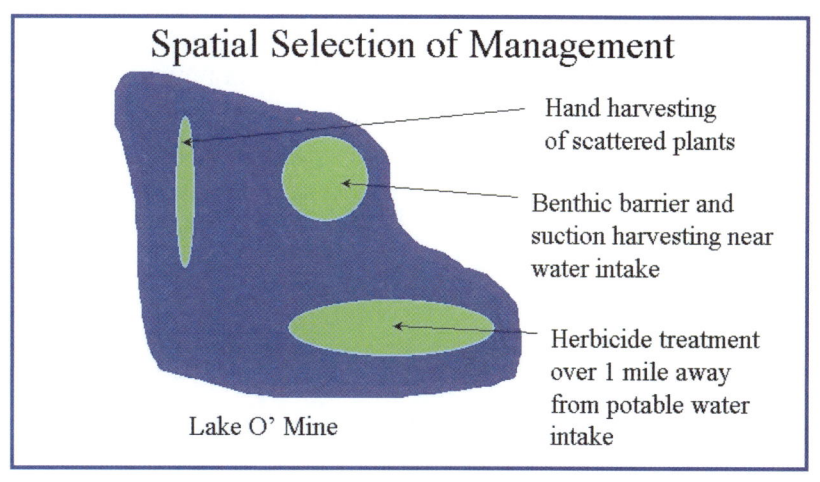

Spatial Selection of Management

- Hand harvesting of scattered plants
- Benthic barrier and suction harvesting near water intake
- Herbicide treatment over 1 mile away from potable water intake

Lake O' Mine

or irrigation water intake. The primary use of the water (i.e., drinking or irrigation) may preclude the use of herbicides that cannot be applied to waters used for drinking or irrigation; therefore, the most appropriate control method for this area might be the use of a benthic barrier and suction harvesting. Consider another site that is more than a mile from the same intake. Weeds at this site could be controlled with herbicides without restrictions on other uses (provided the label specifies use of the herbicide in the area). Perhaps you have an area that is colonized mainly by scattered plants instead of dense stands. If the goal is to eradicate the plant from the water body and you have volunteers at your disposal, hand pulling may be the best method to prevent the formation of dense beds of the weeds.

Management techniques may change over time based on the success (or failure) of the management program. For example, consider Long Lake in Washington State, a small body of water that was dominated by Eurasian watermilfoil (Chapter 13.2) throughout more than 90% of the littoral zone. A whole-lake treatment of fluridone was applied to Long Lake, which reduced the biomass of the weeds by more than 90%. Small remaining beds in the second year were managed with diver-operated suction harvesting, benthic barriers or spot treatment with contact herbicides. By the third or fourth year, routine surveys found only sporadic Eurasian watermilfoil fragments, which were removed by hand harvesting. Similar treatment programs have been successful in other water bodies as well, which demonstrates that

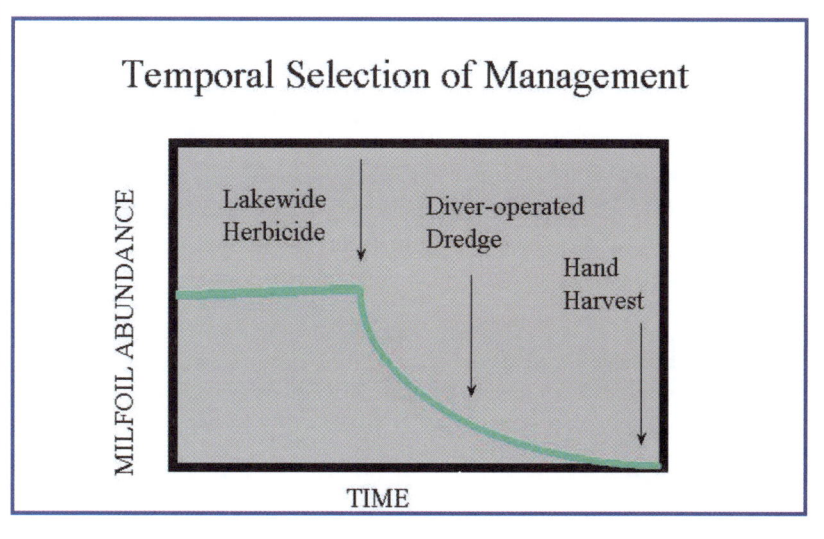

Temporal Selection of Management

MILFOIL ABUNDANCE vs TIME

- Lakewide Herbicide
- Diver-operated Dredge
- Hand Harvest

it is appropriate to alter management techniques as weed control requirements change over time. A wide variety of aquatic plant management techniques may be employed and include physical (Chapter 6), mechanical (Chapter 7), biological (Chapter 8, 9 and 10) and chemical (Chapter 11) control methods. Regardless of method, all techniques should be selected based on their technical merits, as limited by economic and environmental thresholds.

Evaluation

Evaluation of management techniques and programs is typically lacking, even in large-scale management programs. A quantitative assessment should be made to determine the effectiveness of weed management activities, identify environmental impacts (both positive and negative) of management activities, provide the economic cost per acre of management and address stakeholder satisfaction.

Summary

It is critically important to develop a management plan to effectively prevent and control invasive aquatic plants in water resources. Planning should be a continuous process that is ongoing and evolves based on past successes and failures. A comprehensive plan should educate the public about invasive species so they can identify and exclude weeds from uninfested areas. Aquatic plant management programs should also provide a concise assessment of the problem, outline methods and techniques that will be employed to control the weed and clearly define the goals of the program. Mechanisms for monitoring and evaluation should be developed as well and information gathered during these efforts should be used to implement site-specific management and to optimize management efforts. The planning process helps to prepare for the unexpected in weed management, but resource managers should expect the plan to change as stakeholders provide input and management activities commence.

For more information:

- Cover techniques: point intercept (species composition and distribution in the whole lake)
http://el.erdc.usace.army.mil/elpubs/pdf/apcmi-02.pdf
- Cover techniques: line intercept (species composition and distribution in a study plot)
http://el.erdc.usace.army.mil/elpubs/pdf/apcmi-02.pdf
- Abundance techniques: biomass (species composition and abundance)
http://www.hpc.msstate.edu/publications/docs/2007/01/3788JAPM_45_31_34_2007.pdf
- Hydroacoustic techniques: SAVEWS (distribution and abundance; no discrimination among species)
http://www.erdc.usace.army.mil/pls/erdcpub/docs/erdc/images/SAVEWS.pdf
- Remote sensing: satellite, aircraft (distribution of plants near the surface only; no discrimination among species)
http://rsl.gis.umn.edu/Documents/FS7.pdf
- Rockwell HW Jr. 2003. Summary of a survey of the literature on the economic impact of aquatic weeds. Aquatic Ecosystem Restoration Foundation, Lansing, MI.
http://www.aquatics.org/pubs/economic_impact.pdf

Photo and illustration credits

Page 171 upper: Nuisance growth near a water intake; John Madsen, Mississippi State University Geosystems Research Institute

Page 171 lower: Long Lake herbicide treatment; John Madsen, Mississippi State University Geosystems Research Institute

Michael D. Netherland
US Army Engineer Research and Development Center
Gainesville FL
Jeffrey Schardt
FL Fish and Wildlife Conservation Commission
mdnether@ufl.edu; jeff.schardt@myfwc.com

Appendix E

A Manager's Definition of Aquatic Plant Control

Introduction

It would seem like a simple task to define "control", but this appendix illustrates how difficult and variable the term can be. Even scientists argue about the definition of control. For example, entomologists who release a potential biocontrol agent (Chapter 8) may define control as a 10% reduction in plant growth, but most lake managers and homeowners disagree. Do barley straw, enzymes and bacteria really control algae (Chapter 12)? Can native insect populations be augmented to provide weed control? Much depends on the definition of "control."

Defining aquatic plant control

During the past few decades demand for access to and use of US surface waters has increased. These uses include real estate, recreation, irrigation, hydropower, potable water, navigation and efforts to conserve environmental attributes such as fish and wildlife habitat. Aquatic plants are a natural and important component of many freshwater systems and resource managers consider a diverse assemblage and a moderate level of aquatic vegetation to be beneficial for numerous ecosystem functions. Nonetheless, an overabundance of aquatic plants, particularly invasive nonnative plants, can impair freshwater systems, requiring some level of aquatic plant management to conserve water body uses and functions. These aquatic plant management activities routinely take place on water bodies ranging in size from small private ponds to large public multi-purpose lakes and reservoirs.

With increasing demands and values associated with surface waters has come a greater need for aquatic plant control. Nonetheless, the term "control" can take on many meanings depending upon the type and amount of use of each water body, the species of plants present, the responsibilities of resource managers and the objectives of various stakeholder groups associated with the water body. A quick review of reference materials provides the reader with dozens of descriptions and synonyms for "control", and yet for various reasons none provide a meaningful definition for aquatic plant management. The Aquatic Plant Management Society (APMS) has requested that we address this deficiency by providing an aquatic plant manager's working definition of aquatic plant control.

While the terms aquatic plant control and aquatic plant management are often considered synonymous, many resource managers consider control efforts as being operational in nature and management as a process more aligned with program goals and objectives. The APMS defines aquatic plant control as **techniques used alone or in combination that result in a timely, consistent and substantial reduction of a target plant population to levels that alleviate an existing or potential impairment to the uses or functions of the water body.**

This definition best applies to management techniques that directly target a reduction in plant biomass. It is recognized that some management strategies seek to impact factors such as plant reproductive capacity (e.g., production of flowers, seeds, tubers, etc.) or nutrient availability; while these techniques are often recognized as a valuable component of an integrated management program, physical reduction of plant biomass may not result for many years. Moreover, in our definition, the use of the term "substantial" may seem ambiguous; however, we feel there is an inherent problem with using quantitative guidelines (e.g., a 70% reduction in biomass results in acceptable control) to define what is in most cases a series of qualitative field observations by the aquatic resource manager and stakeholders to determine the success of the management activity. Aquatic resource managers should always consider if the proposed management technique has a successful track record and know the limitations of the potential strategy. Claims that a product or technique can provide control should be supported by peer-reviewed literature, experiences from other resource managers with similar management objectives or current research and demonstration efforts.

No single definition of aquatic plant control can cover each specific contingency therefore good communication on the front end is key. The resource manager and stakeholders must first establish expectations for the amount and duration of plant control prior to the initiation of a control activity and then implement a management strategy to meet these expectations. This definition and the following discussion are intended to address factors that relate directly and indirectly to aquatic plant control. Numerous variables influence aquatic plant control operations and many of these parameters, including water body uses, environmental conditions and available management tools, are presented throughout this handbook, along with the influences they may have on the planning or outcomes of aquatic plant control operations. This information may be useful to managers responsible for conserving identified uses and functions of public waterways and who must explain to stakeholders the reasoning behind management plan selection and the ultimate results.

Linking management decisions to aquatic plant control expectations: factors that influence decisions and outcomes

Aquatic plants have been controlled in US surface freshwaters under organized programs for more than a century, so it is natural to ask why it is necessary to provide a definition of aquatic plant control at this point in time. In questioning a number of managers, researchers and other stakeholders, it became obvious that opinions on what constituted acceptable control of an aquatic plant population varied widely. While agricultural managers have been using terms such as "weed free periods" and "crop yield reductions" to define the economic benefits of weed control in cropping systems, aquatic plant managers have a different focus than their terrestrial counterparts. Agricultural weed managers usually attempt to control a broad spectrum of weeds in order to enhance one or more crop species in a fairly controlled environment with a specific function. Aquatic plant managers usually try to control one or two weeds (usually invasive exotic species) to conserve or enhance perhaps dozens of desirable plants as well as multiple uses of aquatic systems. In essence, an agricultural definition of "weed control" does not encompass the issues associated with aquatic plant management.

In developing a manager's definition of control, it was initially tempting to utilize the language of research to provide a quantitative definition. Both the amount and duration of plant control can be readily quantified within the framework of an experimental study or demonstration project. Nonetheless, many experimental studies result in destructive sampling of the target plants at a given point in time (e.g., 90% reduction 8 weeks after treatment) and they often don't allow us to determine if even better control or subsequent recovery would result at a later point in time. While this efficacy information can be very useful to managers regarding the expected performance of a management technique, the uses, functions and environmental conditions can vary widely among water bodies and within water bodies through time. This will influence not only the level of management that may be attempted, but also the outcomes of each control operation. While research projects utilize methods that allow for quantification of control, the vast majority of aquatic plant control operations are ultimately judged by fairly subjective visual observations and qualitative means (e.g. the target plants are near the bottom, difficult to find and the current level of control is rated as good). Therefore, plant control or lack thereof is largely based on whether or not the resource manager and stakeholder expectations have been met.

As noted above, there are numerous issues that either directly or indirectly influence aquatic plant control and management strategies. Before selecting control tools or developing management strategies, three key elements should be addressed that will ultimately influence the manager's decision making process.

Native vs. nonnative vs. invasive aquatic plant control
The National Invasive Species Council defines an invasive species as an alien species whose introduction does or is likely to cause economic or environmental harm or harm to human health. While there are major distinctions between invasive exotic and native species, the main objective of this paper is to clarify the term "control" and as such will not make significant distinctions between managing invasive exotic species and nuisance growth of native plants. Whether a plant is a native or exotic, it can cause problems for given water uses (e.g., water conveyance, access). Nevertheless, two key distinctions between nuisance native and invasive plants deserve further discussion. First, problems associated with nuisance native vegetation are typically site-specific, whereas invasive plants can impair uses and functions of waters across a broad spectrum of conditions and on a regional scale. The vast majority of large-scale aquatic plant control efforts in the US target invasive species. These plants have the potential to spread and dominate new ecosystems and they also have demonstrated the ability to become established in relatively stable aquatic systems. The philosophy behind invasive plant management programs often is to reduce the potential for spread within and among water bodies by reducing the plant biomass to the greatest extent practicable. The second distinction involves early detection and rapid response (EDRR) programs. These efforts are typically unique to invasive exotic species. A significant and costly multi-agency effort may be initiated to control a very small infestation; however, given the potential negative properties of many invasive exotic plant species, these front-end efforts are viewed as necessary and cost-effective.

Efficacy vs. control
It is tempting to define aquatic plant control in terms of an expected percent reduction in coverage or biomass of a target plant population. Some regulatory agencies (e.g., California EPA, Canada Pest Management Regulatory Agency) require that herbicide manufacturers prove the efficacy of their products prior to registration. In this regulatory scenario, a product must reduce a target pest population by greater than 70 or 80% to provide efficacy. Within the discipline of aquatic plant management, numerous techniques can provide both a rapid and significant reduction in a target plant population (>70%), but these results may only be sustained for a few weeks or months. Therefore, depending upon when the efficacy of a management technique is measured, one assessment may suggest that control was achieved, whereas a subsequent assessment conducted weeks, months or a season later may lead to the conclusion that the management effort failed to provide any level of control.

If resource managers and stakeholders have agreed to implement a strategy to provide an entire season of biomass reduction and the target plants recover within one or two months, then by our definition, control has not been achieved. In contrast, some methods may result in slow initial impact on a weed population, but may ultimately provide one or more seasons of control. To complicate matters, many stakeholders fail to grasp that an aquatic plant problem may require more than one treatment or strategy. It is incumbent upon resource managers to understand the strengths and weaknesses of the various management techniques and then convey this information to the stakeholders. If expectations are not defined properly, the stakeholder may lose confidence in the management program. When managers do not establish clear expectations, they are often questioned as to whether control was achieved. Attempting to assess aquatic plant control when clear expectations were not established on the front end is one of the biggest challenges in coming up with a meaningful definition or even assessment of control.

Environmental controls

Managers must be careful not to confuse slow-acting control methods with natural variations in plant populations. While it is often tempting to link a prior control effort with the large-scale decline of a target plant population, environmental events (e.g. droughts, floods, hurricanes, seasonal senescence, etc.) often are largely responsible for these declines. If sufficient data do not exist to support a cause and effect relationship between a control effort and plant biomass decline, managers should avoid making claims that cannot be supported by evidence. Some managers rely on environmental events (e.g. flooding events that scour submersed plants or move floating vegetation; prolonged periods of high, dark water that prevent light penetration for submersed plants) to provide control. While this can be effective, in order to be considered an aquatic plant management technique, there should be some level of predictability associated with the environmental event. From a management perspective there is a big difference in relying on routine seasonal flooding events to control a given plant population versus relying on 100-year floods or droughts to provide plant control.

Levels of aquatic plant control

At the most basic level there are three possible aquatic plant control approaches:

1) no attempt to control,
2) control efforts to eradicate a plant species,
3) some level of intermediate control that is either incomplete or temporary.

No attempt to control

Despite its connotation, the "no control" option is a valid management decision whose potential outcomes must be considered by managers and explained to stakeholders. Factors that influence a manager not taking active control measures may include:

- plant species – Is the plant invasive? Is it a native plant impairing water body uses or is it just unwanted by stakeholders?

- size of infestation – Is this a pioneer infestation consisting of a few plants? Is it an established, but stable, population? Is it an established population or starting to approach problematic thresholds?

- plant location – Is the infestation in an isolated location? Is the location conducive to spreading the pest plant by fragmentation, flow, etc.? Are there important nearby water bodies that are prone to becoming infested?

- plant biology – Is there a likelihood of a rapid population expansion? Would "no control" permit the plant to produce viable seed or vegetative propagules that could make later control efforts more difficult and expensive?

- exploitation – Is the plant species providing an ecological service (e.g. nutrient uptake, food source for waterfowl, habitat for fisheries, etc.)?
- managerial will – Managers may be under pressure to not control a plant because it provides benefits (perceived or real) to a user group. Stakeholders may oppose control because they are not familiar with proposed methods.
- managerial experience – Inexperienced resource managers are often uncomfortable with making aquatic plant management decisions (especially on a large-scale). Until a manager understands the issues and situation, the "no control" option may be viewed as the safest and least controversial.

The consideration of these factors and others may justify a "no control" decision. There are consequences associated with all management decisions and "no control" is not exempt. As previously addressed, plant reductions related to environmental factors could be included within the realm of the "no control" option. While environmental events such as floods, droughts, freezes or severe algae blooms can be quite effective in controlling aquatic plants, these events are not typically predictable and they are not initiated by managers. Nonetheless, the fact that some managers tend to rely on seasonal or weather events to provide effective control suggests the term "no control" may be a misnomer in these situations.

Eradication

Much like defining control, eradication has proven to have numerous meanings to various managers, researchers and stakeholders. In a strict sense, eradication means the complete and permanent removal of all viable propagules of a plant population. This is confounded when a population is removed and then reintroduced at a later time. Some plants may be eradicated following single management efforts [e.g., removal of waterhyacinth (Chapter 13.5) plants prior to seed set], whereas others such as hydrilla (Chapter 13.1) may requires years of intense surveillance and management. Eradication efforts are typically employed when a region, state or watershed is threatened with a new introduction of an invasive species that has potential for significant economic or environmental impact. Based on efforts by various resource management agencies to date, aquatic plant eradication programs are characterized by:

- sustained and multi-year efforts to insure elimination of the plant population
- small-scale efforts to control relatively few plants
- control costs on a per acre basis can be quite high
- the overall impact of repeated control efforts on the infested water body is continually weighed against the regional threat posed by the invasive plant
- control efforts may eventually be reduced; however, vigilant monitoring remains a key to success

Temporary control

Outside the realm of eradication, all other control efforts are temporary. Temporary control is essentially an acknowledgement that 100% control is either not an economically viable management objective or is not possible. Temporary control is a continuum that can be represented by the short-term reduction of target plants following mechanical harvesting or spot treatments with contact herbicides to many years of control that may result from grass carp (Chapter 10) stocking for submersed plants or decades of suppression of alligatorweed by the alligatorweed flea beetle (Chapter 9). Thus, temporary control results when the aquatic plant manager has made the decision that eradication is not a viable endpoint and some level of target plant persistence is acceptable in the management strategy for a given water body.

Temporary control is achievable using a variety of methods. Managers should evaluate each proposed method and the integration of various methods in terms of meeting specific control objectives.

Maintenance control

Maintenance control is applied on a lake-wide or regional scale over time, usually to reduce and contain invasive species. Once established, invasive aquatic plants can be extremely difficult, if not impossible, to eradicate. However, managing invasive plants at some prescribed level that does not impair the uses and functions of the water body can reduce environmental and economic impacts. As the term implies, maintenance control indicates that a conscious decision has been made to actively control an aquatic plant problem with the understanding that a long-term commitment to management rather than eradication is the goal. Simply stated, maintenance control involves routine, recurring control efforts to suppress a problem aquatic plant population at an acceptable level.

Maintenance control encompasses a continuum of control objectives. On one extreme, the goal of maintenance control may be to reduce and sustain a plant population at the lowest feasible level that technology, finances and conditions will allow. This strategy has proven effective in managing established populations of highly invasive aquatic plants. By managing waterhyacinth at low levels through frequent small-scale control operations, there is a corresponding reduction in the overall management effort, especially herbicide use and management costs. There also are environmental gains, such as reductions in sedimentation and dissolved oxygen depressions. At the other end of the spectrum, maintenance control operations can be applied just prior to plant populations impairing the uses or functions of the water body. This strategy entails allowing plants to grow to the brink of problem levels and therefore may be best employed to control slow-growing or otherwise non-invasive plants.

Paradoxically, there is often more stakeholder support for crisis management (allowing plants to reach some problem or impairment level) than maintaining invasive species at low levels. This may be related to stakeholders being unaware of invasive plant growth potential. It also may be related to the public's perceptions of control methods – for example, not understanding that less herbicide may be needed to maintain plants at low levels rather than waiting for an obvious problem to develop.

Adaptive management

Since maintenance control represents a long-term commitment, it must also encompass a strategy known as adaptive management. Uses and functions of water bodies change through time, as do conditions within water bodies and among plant populations. Examples include target and nontarget plant growth stages, water temperature, depth, clarity and flow. All change several times during the year and can require different control strategies or different expectations for control outcomes. Therefore, integrated management plans for each aquatic plant control operation must account for and adapt to these changes.

Communicating control expectations to user groups

Many stakeholders view aquatic plant management endeavors as a one-time control effort with no further need for additional management. This does not reflect the reality of the discipline of aquatic plant management. The vast majority of management programs require a sustained effort over multiple years to keep unwanted vegetation under control. For example, while grass carp can provide long-term control of hydrilla, this result is due to their continuous presence and feeding on any plant regrowth. Carp can sustain control for many years, yet removal of the carp due to natural losses or on purpose will typically result in the recovery of the target plant. Likewise, a single treatment with the herbicide fluridone (Chapter 11) may remove a target invasive plant such as Eurasian watermilfoil (Chapter 13.2) within a system for one to several years. Upon discovery of new plants, many stakeholders are dismayed that the treatment did not eradicate the problem. In some cases these plants may have regrown from seed or they may have been introduced from a nearby lake or reservoir that was not managed. Aside from the use of an effective classical biological control organism (highly selective – Chapter 8) or high stocking rates of grass carp (non-selective – Chapter 10), user groups must be informed about the importance of maintaining continuity in an aquatic plant management program. Single small-scale efforts that don't address the problem at an adequate scale often lead to claims that "we tried that and it didn't work." A lake full of hydrilla or Eurasian watermilfoil may require whole-lake management efforts. The control may last one, two or more seasons, but experience suggests that these invasive plants will ultimately return.

One of the bigger challenges facing aquatic resource managers relates to the promotion of unproven and often costly technologies that are packaged as environmentally friendly approaches to aquatic plant management. As noted earlier, claims of a product or device providing "control" should be supported by published or ongoing research or by another reputable resource manager who has successfully applied that technique or strategy and met similar control objectives.

Appendix E

Appendix F

Miscellaneous Information

Common names, trade names, formulations and registrants or suppliers of aquatic herbicides and algicides. Labels, MSDS and other product information is available on the websites of most registrants or suppliers. This is not a complete listing of all products that are registered for aquatic use. The mention of a trade or brand name does not constitute an endorsement of the product by the authors, editors or AERF.

Copper chelates – ethanolamines		
Trade name	Formulation	Registrant or supplier
Captain	Liquid	SePRO
Clearigate	Liquid	Applied Biochemists
Cutrine–Plus	Liquid	Applied Biochemists
Cutrine–Plus	Granular	Applied Biochemists
Cutrine–Ultra	Liquid	Applied Biochemists

Copper chelates – ethylene diamine		
Trade name	Formulation	Registrant or supplier
Harpoon	Liquid	Applied Biochemists
Komeen	Liquid	SePRO

Copper chelates – triethanolamine		
Trade name	Formulation	Registrant or supplier
K–Tea	Liquid	SePRO
Symmetry	Liquid	Phoenix

Copper chelates – triethanolamine + ethylene diamine		
Trade name	Formulation	Registrant or supplier
Nautique	Liquid	SePRO

Copper citrate/gluconate		
Trade name	Formulation	Registrant or supplier
Algimycin–PWF	Liquid	Applied Biochemists

Copper sulfate		
Trade name	Formulation	Registrant or supplier
AB Brand	Crystals	Applied Biochemists
ChemOne Copper	Crystals	ChemOne
Current	Liquid	Phoenix
EarthTec	Liquid	Earth Science Laboratories
Formula F–30	Liquid	Diversified Waterscapes
Old Bridge Copper Sulfate	Crystals	Old Bridge Chemicals
Triangle Brand	Crystals	Freeport–McMoRan

Diquat		
Trade name	Formulation	Registrant or supplier
Diquat E–Pro 2L	Liquid	Nufarm Americas
Harvester Landscape and Aquatic	Liquid	Applied Biochemists
Helm Diquat Aquatic and Landscape	Liquid	Helm Agro US
Littora	Liquid	SePRO
Redwing	Liquid	Phoenix
Reward	Liquid	Syngenta
Weedplex Pro Landscape and Aquatic	Liquid	Sanco Industries

Dyes—Acid Blue 9 and Tartrazine		
Trade name	Formulation	Registrant or supplier
Admiral	Liquid	Becker Underwood
Aquashade	Liquid	Applied Biochemists

Endothall—amine salt		
Trade name	Formulation	Registrant or supplier
Hydrothol 191	Liquid	United Phosphorus
Hydrothol 191	Granular	United Phosphorus

Endothall—potassium salt		
Trade name	Formulation	Registrant or supplier
Aquathol	Liquid	United Phosphorus
Aquathol Super–K	Granular	United Phosphorus

Fluridone

Trade name	Formulation	Registrant or supplier
Avast SC	Liquid	SePRO
Sonar AS	Liquid	SePRO
Sonar One	Granular	SePRO
Sonar PR	Granular	SePRO
Sonar Q	Granular	SePRO
Sonar SRP	Granular	SePRO
Whitecap	Liquid	Phoenix

Glyphosate

Trade name	Formulation	Registrant or supplier
Aquamaster	Liquid	Monsanto
AquaNeat	Liquid	Nufarm Americas
AquaPro	Liquid	SePRO
Avocet	Liquid	Phoenix
Avocet PLX	Liquid	Phoenix
Rodeo	Liquid	Dow AgroSciences
Shore Klear	Liquid	Applied Biochemists
Touchdown Pro	Liquid	Syngenta

Hydrogen peroxide

Trade name	Formulation	Registrant or supplier
Greenclean	Liquid	Biosafe Systems
Greenclean Pro	Granular	Biosafe Systems
PAK–27	Granular	Peroxygen Solutions
Phycomycin SCP	Granular	Applied Biochemists

Imazamox

Trade name	Formulation	Registrant or supplier
Clearcast	Liquid	BASF

Imazapyr

Trade name	Formulation	Registrant or supplier
Ecomazapyr	Liquid	Alligare
Gullwing	Liquid	Phoenix
Habitat	Liquid	BASF
Polaris	Liquid	Nufarm Americas

Penoxsulam

Trade name	Formulation	Registrant or supplier
Galleon	Liquid	SePRO

Triclopyr acid

Trade name	Formulation	Registrant or supplier
Trycera	Liquid	Helena

Triclopyr amine

Trade name	Formulation	Registrant or supplier
Ecotriclopyr	Liquid	Alligare
Kraken	Liquid	Phoenix
Navitrol DPF	Granular	Applied Biochemists
Renovate	Liquid	SePRO
Renovate 3 OTF	Granular	SePRO
Tahoe 3A	Liquid	Nufarm Americas

Triclopyr + 2,4–D amine

Trade name	Formulation	Registrant or supplier
Renovate Max G	Granular	SePRO

2,4–D acid

Trade name	Formulation	Registrant or supplier
Hardball	Liquid	Helena
Sinkerball	Liquid	Helena

2,4–D amine

Trade name	Formulation	Registrant or supplier
DMA 4 IVM	Liquid	Dow AgroSciences
Sculpin	SePRO	Applied Biochemists
Weedar 64	Liquid	Nufarm Americas
Weedestroy AM40	Liquid	Nufarm Americas

2,4–D butoxy-ethyl ester

Trade name	Formulation	Registrant or supplier
Navigate	Granular	Applied Biochemists

For more information

Alligare LLC. http://www.alligarellc.com/

Applied Biochemists. http://www.appliedbiochemists.com/

BASF—The Chemical Company. http://www.basf.com/

Becker Underwood. http://www.beckerunderwood.com/

Biosafe Systems LLC. http://www.biosafesystems.com/

ChemOne Inc. http://www.chemone.com/

Diversified Waterscapes. http://www.dwiwater.com/

Dow AgroSciences. http://www.dowagro.com/

Earth Science Laboratories Inc. http://www.earthsciencelabs.com/

Freeport–McMoRan Copper and Gold Inc. http://www.fcx.com/

Helena Chemical Company. http://www.helenachemical.com/

Helm Agro US Inc. http://www.helmagro.com/

Monsanto Company. http://www.monsanto.com/

Nufarm Americas Inc. http://www.nufarm.com/

Old Bridge Chemicals Inc. http://www.oldbridgechem.com/

Peroxygen Solutions. http://www.peroxygensolutions.com/

Phoenix Environmental Care. http://www.phoenixenvcare.com/

Sanco Industries Inc. http://www.sancoind.com/

SePRO Corporation. http://www.sepro.com/

Syngenta Global. http://www.syngenta.com/

United Phosphorus Inc. http://www.upi-usa.com/

Common Conversion Factors (David Petty, NDR Research)

To change	To	Multiply by
acres	hectares	0.4047
acres	square feet	43,560
centimeters	inches	0.3937
centimeters	feet	0.03281
cubic feet	cubic meters	0.0283
cups	ounces (liquid)	8
cubic meters	cubic feet	35.3145
cubic meters	cubic yards	1.3079
feet/second	miles/hour	0.6818
gallons (U.S.)	liters	3.7853
grams	ounces (avdp)	0.0353
grams	pounds	0.002205
hectares	acres	2.471
inches	centimeters	2.54
kilograms	pounds (avdp)	2.2046
kilometers	miles	0.6214
liters	gallons (U.S.)	0.2642
liters	pints (liquid)	2.1134
liters	quarts (liquid)	1.0567
meters	feet	3.2808
meters	yards	1.0936
miles	kilometers	1.6093
miles	feet	5280
miles/hour	feet/minute	88
ounces (avdp)	grams	28.3495
ounces (avdp)	pounds	0.0625
ounces (liquid)	pints (liquid)	0.0625
ounces (liquid)	quarts (liquid)	0.03125
pints (liquid)	liters	0.4732
pints (liquid)	ounces (liquid)	16
pounds (avdp)	kilograms	0.4536
pounds	ounces	16
quarts (liquid)	liters	0.9463
quarts (liquid)	ounces (liquid)	32
square feet	square meters	0.0929
square kilometers	square miles	0.3861
square meters	square feet	10.7639
yards	meters	0.9144
1 ppm	= 1 mg/L or 1 mg/kg	
1 ppb	= 1 µg/L or 1 µg/kg	

Common Water Quality Parameters (David Petty, NDR Research)

Alkalinity: The water's ability to neutralize acids, measured in milligrams per liter of total alkalinity as equivalent calcium carbonate (mg/L $CaCO_3$). Alkalinity helps regulate pH and metal content in water. Levels of 20-200 mg/L are common in fresh water systems.

Conductivity: The measure of the capacity of water to conduct an electric current, measured in either microSiemens per centimeter of water at 25 degrees centigrade (µS/cm @25°C) or micromhos per centimeter (µmhos/cm). Conductivity is an indirect measure of dissolved solids such as chloride, nitrate, sulfate, phosphate, sodium, magnesium, calcium, and iron.

Dissolved oxygen (DO): The amount of oxygen measured in water in milligrams per liter (mg/L). In general, rapidly moving water contains more dissolved oxygen than slow or stagnant water and colder water contains more dissolved oxygen than warmer water. Low DO levels can lead to fish kills. Optimal DO for many species is between 7 and 9 mg/L.

Hardness: Water hardness is generally the measure of the cations of magnesium and calcium in the water, usually expressed as mg/L. Waters with a total hardness in the range of 0 to 60 mg/L are termed soft; from 60 to 120 mg/L moderately hard; from 120 to 180 mg/L hard; and above 180 mg/L very hard.

pH: Scale of values from 0 to 14 which indicate the acidity of a waterbody. Water is acidic if pH is below 7, with increasing acidity with lower values. Water is basic when above 7, and more basic with increasing values. A value of 7 is considered neutral pH. Aquatic organisms differ in the pH range they can tolerate and flourish in.

Turbidity: A measure of the amount of particulate matter that is suspended in water, and is measured in Nephelometric Turbidity Units (NTU). Water that has high turbidity appears cloudy or opaque. High turbidity can cause increased water temperatures because suspended particles absorb more heat and can also reduce the amount of light penetrating the water.

Appendix F

Glossary of terms

Note: words in this glossary are defined in the context in which they are used in this manual

A

Abscission: a process in which part of a plant naturally detaches from the rest of the plant

Absorb: to soak up a substance

Acidic: having a pH of less than 7; compare to alkaline

Acre: an area containing 43,560 square feet

Acre-foot: the amount of water one foot deep in an area that covers one acre; equal to 325,851 gallons of water with a weight of approximately 2.7 million pounds; used to calculate the amount of herbicide to be applied to a body of water

Active ingredient: the specific chemical that has herbicidal activity and is responsible for killing or controlling a plant

Acute: severe or sharp, as in the shape of a leaf; or meaning rapid or quick when referring to toxicity

Adsorb: to bind to the outside or surface, such as herbicides binding to soil particles

Adsorption: the adhesion or accumulation of a substance onto another, such as herbicides binding to soil particles

Adventive: a nonnative organism that colonized an area long ago, developed a reproducing population and has become naturalized

Aeration: the introduction of oxygen to water, often accomplished with an aerator

Aerobic: containing oxygen; compare to anaerobic

Alkaline: having a pH of greater than 7; also called basic; compare to acidic

Allocation: distribution of a substance to different areas within an organism

Amphibian: an air-breathing organism that can live in terrestrial and aquatic environments

Amphipod: a small crustacean often eaten by juvenile fish

Anaerobic: lacking oxygen, as in some highly organic lake sediments; compare to aerobic

Annual: a plant that completes its entire life cycle in one year or season; compare to perennial

Anthropogenic: occurring as a result of human activity

Apical bud, apical meristem: a growing point in the uppermost portion of many plants

Arthropod: an invertebrate organism with a segmented body; examples include insects and crustaceans

Augmentation: a process where additional organisms are added to supplement existing populations; used in biocontrol

Auxin: a plant hormone that regulates growth

Axil: the area where the leaf stalk or petiole attaches to the stem

Axillary bud, lateral bud: a meristem or bud in the leaf axil or along the sides of stems; compare to apical bud

B

Ballast: weight, typically in the form of water, placed into the hull of a heavily loaded cargo ship to increase stability; usually removed or discharged when cargo is removed

Basic: see alkaline

Bathymetry: the measurement of water depths within a body of water

Benthic: relating to the bottom of a water body and the organisms that inhabit the sediments

Bioaccumulation: a process where a substance builds up in an organism after the organism consumes other organisms contaminated with the substance

Bioconcentration: the buildup of a substance in an organism at levels greater than the surrounding environment

Biocontrol: the use of an organism such as an insect or fish to control an invasive organism such as an aquatic weed

Biodiversity: a measure of the number of different species in an environment

Biomass: the amount of vegetative material (leaves, stems, etc.) produced by a plant

Biotype: an organism that differs (in appearance or another characteristic) from other organisms of the same species; sometimes referred to as a variety or subspecies

Brackish: a mixture of fresh and saline water

Bulblet: a bulb-like vegetative structure produced by some plants that is capable of forming a new plant

Bycatch: the unintentional trapping of organisms during mechanical harvesting of aquatic weeds

C

Calcified: the accumulation of calcium deposits on the leaves of a plant

Chelate: an organic compound which binds with ions such as copper

Chlorophyll: the green pigment in plants and other photosynthetic organisms that use light to produce energy

Chloroplasts: plant structures where sunlight is converted to energy

Clarity: the relative clearness of water; usually measured with a Secchi disk; compare to turbidity

Clones: organisms that are genetically identical to one another

Coevolution, coevolved: a process where different organisms in the same environment evolve or change in concert; for example, insects and plants that have evolved together over time to provide services to one another

Crown: the region of a plant where the stems and the root join together

Crustacean: an aquatic arthropod with a segmented body and hard exoskeleton; examples include lobsters, shrimp and crabs

Cuticle: a protective waxy layer that is present on the leaves of terrestrial plants but absent on the leaves of most submersed aquatic plants

Cyanobacteria: photosynthetic bacteria; also call blue-green algae

D

Deactivation: a process where a substance is rendered inactive due to a process within a plant or binding with the sediment

Defoliation: loss or removal of a plant's leaves

Degradation: breakdown of complex organic compounds into simpler substances that are then further degraded or broken down

Depuration: cleansing or purification

Desiccate: to dry out by removing most or all water from an organism

Destratification: loss of the layering that occurs in bodies of water (usually during the summer) and results in water mixing across depths within a water body; see thermocline

Detritivores: organisms that eat detritus or other dead organic matter

Detritus: decomposed organic material (primarily dead aquatic plants) that settles on and in the sediment

Dewatering: the process of removing the water from an aquatic system; see drawdown

Dicotyledon (dicot): a plant characterized by having two seed leaves at germination and leaf veins that are arranged like a net; most broad-leaved plants are dicots; compare to monocotyledon

Diluent: a substance (usually water) used to reduce the concentration of a herbicide and to facilitate uniform application

Dioecious: a condition where individual plants bear only staminate (male) or pistillate (female) flowers; compare to monoecious

Diploid: an organism with two sets of chromosomes; usually fully fertile and able to reproduce by sexual means

Dissipation: the slow reduction in concentration and eventual loss of a substance through degradation, dilution or both processes

Dormant: a condition where plants cease growth in order to survive adverse conditions and resume growth when conditions improve

Drawdown: partial or complete removal of the water in an aquatic system for a period ranging from several months to several years to cause desiccation and death of aquatic weeds

Dredge: removal of part of the sediment in a water body to improve navigation and/or control aquatic weeds; also used to describe the equipment used in this process

E

Ecosystem: the flora, fauna and environmental conditions within a given area

Efficacy: effectiveness

Embayment: a bay-shaped indentation in the shoreline that is larger than a cove but smaller than a gulf

Emergent: a plant that is rooted in the sediment with most parts of the plant maintained above the waterline; examples include most shoreline plants such as cattail, purple loosestrife and pickerelweed

Emulsifier: a substance that is used to keep particles in solution in a fluid; often added to concentrated herbicides so they can be mixed with water

Endemic: considered native or naturally occurring in an area

End-use product: the final product purchased by applicators; usually manufactured with technical grade active ingredients and diluted with inert ingredients such as water and emulsifiers to make the product easy to dilute and apply

Entomology: the study of insects

Enzyme: a chemical that degrades or breaks down a substance or allows a chemical reaction to occur

Equilibrium: a balanced system with little change in the elements that comprise the system

Eradication: complete elimination of an organism from a system; see extirpated

Estuary: the wide part of a river where it nears the ocean

Eutrophic: rich in minerals and organic nutrients; eutrophic conditions encourage algae growth and reduce levels of dissolved oxygen

Eutrophication: the accumulation of excessive minerals and organic nutrients

Evergreen: a plant that maintains its leaves and sometimes continues to grow throughout the year

Exotic: not native to a region or system

Extirpated: see eradication

F

Fauna: collectively, the animals (including insects) present in a system

Floating–leaved: a plant that is rooted in the sediment and has leaves that float on the surface of the water; examples include waterlily and waterchestnut

Flora: collectively, the plants present in a system

Formulation: the form in which a herbicide is sold (liquid, granular or other form)

Fragmentation: a process whereby part of a plant is removed from the rest of the plant due to natural (see abscission) or mechanical means

Free-floating: a plant with roots that typically occupy the upper portion of the water column; examples include waterhyacinth and salvinia

G

Generalist: an organism that does not require a specific food source for growth, survival and reproduction

Genotype: the genetic composition of an organism

Genus: a classification that describes a group of closely related organisms; each genus is further divided into species, whose members are very closely related and can breed with one another

Geotextile: a specialized fabric-like material used to stabilize shorelines or to smother submersed aquatic weeds

GLP: an acronym for good laboratory practices, a set of protocols that must be followed when testing herbicides

H

Half-life: the period of time required for the concentration of a chemical to be reduced by half, usually by microbes, light or chemical reactions

Hardness: a measure of the amount of calcium and carbonates in water

Herbaceous: a "fleshy" plant with no little or no woody material

Heterogeneity: a measure of the genetic diversity in an organism; also used to describe diverse plant communities

Heterotypic: of a different form or type

Hydrology: the study of the properties, distribution and effects of water on the earth's surface, soil and atmosphere

Hydrolysis: the splitting of a compound into two smaller parts as a result of contact with water

Hydropower: energy derived from the force of moving water

Hypereutrophic: extremely high in nutrients; characterized by excessive algae growth that causes water to be very cloudy with poor transparency

Hypolimnetic: pertaining to the hypolimnion, the cold deeper area of a stratified lake

I

Inactivation: a process where a substance is rendered inactive due to a process within a plant or binding with the sediment

Indigenous: native to a region or system

Inert: a substance that lacks herbicidal properties

Inflorescence: the structure and arrangement of a plant's flowers

Insectivorous: insect-eating

Inundated: flooded or under water

Invasive: a species that steals resources from desirable species and reduces diversity by being more competitive than other organisms in the system; most invasive species are nonnative, fast-growing and lack natural enemies

Invertebrate: an animal that lacks a backbone

L

LC50: abbreviation for lethal concentration 50%; the external or applied concentration of a substance required to cause death in 50% of the organisms tested; similar to LD50 (lethal dose 50%)

Larvae: early stage of insect development; examples include maggots and grubs

Lateral: a bud or branch produced from a leaf axil or other non-terminal bud on the plant

Limnology: the study of freshwater systems, including lakes, rivers and ponds

Littoral: the zone near the shoreline where water is typically shallow; usually inhabited by aquatic plants

M

Macrophyte: a plant that can be easily seen without magnification

Macroscopic: an organism that can be easily seen without magnification

Meristem: the part of a plant from which new growth originates; also called a bud

Mesotrophic: having moderate amounts of nutrients and phytoplankton

Metabolite: the product resulting from chemical breakdown or degradation of a more complex organic molecule

Microbe: a tiny organism such as a bacterium or fungus; also called microorganism

Microcrustaceans: very small zooplankton or crustaceans that feed on phytoplankton and are not easily viewed without a microscope or magnifying lens

Microfauna: animals that are not easily viewed without a microscope or magnifying lens

Micronutrient: an element that organisms require in small quantities for healthy growth

Midrib: the central vein of a leaf

Mineralization: the conversion of an element from an organic form to an inorganic form as a result of microbial decomposition

Molting: the shedding of an insect's outer layer to allow expansion and growth

Monocotyledon (monocot): a plant characterized by having a single seed leaf at germination and leaf veins that are arranged in a parallel manner; grasses are monocots; compare to dicotyledon

Monoculture: a group of plants consisting solely of members of a single species

Monoecious: a condition where individual plants bear both staminate (male) and pistillate (female) flowers; compare with dioecious

Monotypic: composed of organisms of the same type or species

Morphology: the appearance of an organism

N

Native range: the geographic region from which an organism originates

Naturalized: a nonnative organism that reproduces and maintains a population in a new area; see adventive

Niche: a specific range of environmental conditions or a habitat in which a species can thrive

Nonindigenous: a nonnative organism

Nutlet: a small, hard, reproductive structure

O

Obligate: requiring a certain environment or food source to survive, grow and reproduce

Off-patent: a chemical that is no longer protected by a patent and can be produced by other companies in addition to the company that developed the product; often available in generic form

Oligotrophic: very low in minerals and organic nutrients

Omnivorous: consuming almost any type of plant or animal matter

Ornithology: the study of birds

Outcompete: make better or more efficient use of available resources than other organisms; deplete resources needed for growth of other organisms

Overwinter: to survive throughout the winter, often in a dormant state or as a propagule

Oxbow: a sharp, U-shaped bend in a river that is no longer attached to the river

Oxygen: present in water at concentrations ranging from 0 to 15 ppm; few fish can survive extended periods when oxygen content is below 2 ppm

Oxygenation: to increase the oxygen content of water, usually with the introduction of air into the system; see aeration

P

Palmate: arrangement where leaflets (small leaves) radiate from a central point; similar to fingers radiating from the palm of the hand

Parasite: an organism that survives by feeding on, damaging or deriving nutrients from another organism

Pathogen: an organism that causes disease to another organism

Pathology: the study of pathogens

Pelagic: referring to deep, cold water; see hypolimnion

Perennial: a plant that requires multiple years or seasons to complete its entire life cycle; compare to annual

Petiole: the "stalk" attaching a leaf to the stem of a plant

Photolysis, photolytic: the breakdown or chemical decomposition of a compound induced by light

Photosynthesis: the daytime-only process by which plants use carbon dioxide to convert sunlight into energy and oxygen

Phytoplankton: tiny, free-floating photosynthetic aquatic organisms; examples include diatoms, dinoflagellates and some species of algae

Pigment: a substance that produces a distinct color in a plant; may have protective properties

Pinnate: resembling or arranged like a feather

Piscivorous: fish-eating

Pistillate: a flower bearing female reproductive structures and lacking male reproductive structures; compare to staminate

Plankton: very small free-floating aquatic organisms; examples include phytoplankton and zooplankton

ppb: parts per billion (1 in 1,000,000,000)

ppm: parts per million (1 in 1,000,000)

Precipitate: settle as a solid to the bottom of the water body

Precipitation: a chemical reaction or process that reduces the solubility of a substance and causes it to precipitate

Predation: consumption of an organism (prey) by another organism (predator)

Pristine: natural; not affected by human activity

Productivity: the trophic state of a lake (biological productivity) or the amount of organic matter produced (plant productivity)

Propagation: the act of creating new plants through sexual or vegetative means

Propagules: vegetative or sexual structures with the ability to create new plants; examples include turions, tubers, bulblets, fragments, winter buds and seeds

Protozoan: a single-celled microscopic organism; examples include amoebas and ciliates

Psyllid: an insect in the family Psyllidae; also called jumping plant lice

Pupa: the stage in insect development between larva and adult; pupae are usually protected within a hard cocoon or case

Q

Quiescence: a resting state

R

Ramet: a new plantlet formed by vegetative means; often borne on a runner or stolon

Recolonization: the re-establishment of a species that was previously found in a system but disappeared

Registrant: the organization responsible for the registration of a pesticide with the US EPA

Reservoir: a man-made body of water used for water storage, flood control, hydropower, recreation or other anthropogenic activities

Residue: any substance in food, water or an organism that occurs as a result of application of a pesticide

Resistant: the ability of an organism to survive or be unaffected by a stressor such as a herbicide; compare to susceptible

Respiration: a process in which plants take up oxygen and release carbon dioxide

Rhizome: modified plant structure that grows underground and has buds that can produce new plants

Richness: the number of distinct species present in a system

Riparian: relating to the bank or shoreline of a body of water

Rootstock: the roots, crown and rhizomes of a plant

Rosette: plant growth form where leaves radiate from a central point or crown instead of being attached to a stem

Runner: see stolon

S

Salinity: measure of the amount of salt in water

Scour: to clear a channel or remove sediment as a result of wave action, current or flow

Secchi disk: a circular disk divided into black and white sections and used to measure water clarity or transparency

Sediment: the soil or organic material at the bottom of the water body

Sedimentation: the process of accumulating sediment, usually as a result of wave action, erosion, reduced water flow in plant beds or decaying plant material

Seedbank: seeds that fall to the sediment and provide a source for new plants in future seasons

Selective: a herbicide that controls certain plants while leaving others unharmed

Senescence: plant death

Serrated: with toothed margins similar to the blade of a saw

Shoots: upright plant stems

Short-day: a condition where daylength is less than 12 hours in length (winter in the US)

Species richness: the number of different plant or animal species in a defined area

Specificity: the ability of a herbicide to selectively control target plants without causing significant damage to nontarget plants

Spores: reproductive structures produced by ferns such as salvinia

Stamen: the pollen-bearing male reproductive structure of a flower

Staminate: a flower bearing male reproductive structures and lacking female reproductive structures; compare to pistillate

Stolon: a stem-like structure or shoot that creeps along the surface of the soil or sediment; also called runner

Stratification: a layered configuration within a body of water whereby distinct and separate upper (epilimnion), middle (metalimnion) and lower (hypolimnion) layers are evident

Structure: referring to the array of architectures provided by different plants, logs, brush piles and rocks in fish habitats

Submersed: a plant that grows mostly or entirely under water

Subspecies: a division within a species to designate a group of plants that differ substantially from other members of the species

Substrate: see sediment

Surfactant: short for "surface-active agent"; a detergent-like substance that reduces surface tension and increases herbicide coverage and penetration into plant stems and leaves

Susceptible: an organism that is damaged or killed by a stressor such as a herbicide; compare to resistant

Systemic: a substance that moves throughout an organism via translocation through vessels in plants

T

Tannins: acidic yellow to brown substances derived from plant materials such as tree bark, roots, leaves and tea

Taxonomy: a system used to categorize, describe and identify organisms

Technical grade: the purest, most concentrated form of an active ingredient

Temperate: a climate that is warm in the summer and cold in the winter

Terrestrial: not flooded or inundated

Thermocline: the metalimnion or center layer of water in a stratified lake; the most extreme temperature changes occur in the thermocline as opposed to the upper (epilimnion) and lower (hypolimnion) layers

Topped-out: a phenomenon where submersed plants such as hydrilla reach the surface of the water and form dense mats or canopies that reduce penetration of light and oxygen

Toxicant: a substance used to damage or kill an organism

Translocation: an active process of movement of a substance within and throughout the vessels of a plant

Triploid: an organism with three sets of chromosomes; usually sterile and unable to reproduce by sexual means

Trophic: related to nutrition and nutrient levels; productivity

Tuber: a vegetative propagule produced in the sediment to facilitate reproduction and overwintering

Turbidity: the degree to which water clarity is reduced by suspended particles, tannins, algae and other substances; compare to clarity

Turion: a propagule produced in the leaf axils or compressed apical buds of hydrilla to facilitate vegetative reproduction, overwintering, survival and spread

U

Upland: see terrestrial

V

Variety: a division within a species to designate a group of plants that differ substantially from other members of the species; similar to subspecies

Vascular plant: plant with a specialized internal transport or vessel system; sugars are transported in the phloem, whereas water and nutrients are transported in the xylem

Vector: an organism that transmits a disease-causing pathogen

Veliger: snail larvae

W

Watershed: the entire drainage area of a river or the catchment area of lakes

Wetland: an area that is inundated or saturated for long enough periods to support plants that are adapted to living under saturated soil conditions

Whorled: with leaves arranged in groups of three or more at a node

Winter bud: compressed apical bud; similar to turion

Z

Zonation: the separation of areas within an ecosystem into specific zones, with each zone having distinct characteristics that distinguish it from other zones

Zooplankton: microscopic aquatic animals and larvae which usually feed on phytoplankton

Aquatic Ecosystem Restoration FOUNDATION

Almost twelve years ago a group of companies formed a nonprofit foundation to address increasing problems with invasive aquatic weeds in complex, multiple-use ecosystems.

The mission of the AERF is to support research and development which provides strategies and techniques for the environmentally and scientifically sound management, conservation and restoration of aquatic ecosystems. Our research provides the basis for the effective control of nuisance and invasive aquatic and wetland plants and algae. Broad strategic goals include:

1. Providing information to the public on the benefits of conserving aquatic ecosystems. This involves various operationally sound methods which are appropriate for a particular water body to achieve the objectives of a sound management plan. This includes the appropriate use of EPA registered aquatic herbicides and algicides. The foundation has produced Biology and Control of Aquatic Plants: A Best Management Practices Manual, which has become one of the most widely read and used references in the aquatic plant management community. This document can be downloaded from our web site (www.aquatics.org), and illustrates the various ways that aquatic plants can be managed – biological, mechanical, physical, chemical, etc.

2. Providing information and resources to assist regulatory agencies and other entities making decisions that impact aquatic plant management. This goal is partially accomplished by providing independent experts on request to address specifically defined issues. Similarly, AERF has sponsored seminars and symposia throughout the United States on aquatic plant management issues. AERF also assists state and local agencies by providing travel grants for regulatory personnel to participate in aquatic-related professional meetings.

3. Funding research in applied aquatic plant management. AERF has funded ecosystem-related research by independent scientists and graduate students in 20 universities in the United States and with the US Army Corps of Engineers. AERF also promotes the attendance of students at aquatic-related professional meetings by providing assistantships and travel grants to dozens of students annually.

Funding is generated through contributions, membership donations and grants. The operation of the Foundation is managed by an Executive Director. A Board of Directors, composed of donors, provides guidance on the development of annual objectives and the management of fiscal resources. Decisions are made by the Executive Director, such as the selection of subject matter experts, speakers for symposia, AERF participation in seminars and meetings and similar activities that fall within the objectives of the Foundation. A Technical Advisory Committee composed of PhD researchers comments on the soundness of the science in the research proposals and consistency in terms of the Foundation's mission statement.

Carlton Layne, Executive Director
Aquatic Ecosystem Restoration Foundation
clayne@aquatics.org • www.aquatics.org

Your AERF membership is key to:

- ▶ maintaining critical efforts in education and outreach
- ▶ expanding partnerships with regulatory agencies
- ▶ building partnerships
- ▶ supporting high quality research
- ▶ attracting graduate students
- ▶ expanding an already diverse membership
- ▶ being a source for resource management

To donate, join or renew your membership in the AERF please send the completed application form and payment to **Treasurer, AERF, 1860 Bagwell Street, Flint, MI 48503-4406**

Please use the following as a guide in the selection of the desired level of membership. Of course, you are welcome to join AERF at any level and additional donations are appreciated.

Date: _____ Name: _____ Company: _____

Address: _____

Phone: _____ Fax: _____ Email: _____

Web Address: _____

☐ Check here if you would like to receive a free copy of the book with your membership.

☐ Gold is recommended for manufacturers and registrants	$15,000
☐ Silver and above is recommended for formulators	$ 5,000
☐ Bronze and above is recommended for distributors	$ 2,500
☐ Affiliate and above is recommended for consultant and application companies, equipment manufacturers/resellers and biological producers/resellers	$ 1,000
☐ Associate and above is recommended for societies, federal and state agencies, aquatic resource management associations, applicators and consultants	$ 250
☐ Individual and above is recommended for individual members	$ 50
☐ Student and above is recommended for students	$ 0

For more information contact:
Carlton R. Layne, Executive Director
3272 Sherman Ridge Drive, Marietta, GA 30064
Phone: 678-773-1364 • Fax 770-499-0158 • Email clayne@aquatics.org
www.aquatics.org